Albert B. Hussong

A Store of Knowledge

Albert B. Hussong

A Store of Knowledge

ISBN/EAN: 9783337248475

Printed in Europe, USA, Canada, Australia, Japan

Cover: Foto ©berggeist007 / pixelio.de

More available books at **www.hansebooks.com**

A Store of Knowledge

—EMBRACING—

THE EARLY PRINCIPLES OF OUR GOVERNMENT,

TOGETHER WITH ALL THE

Cabinets, and a Brief Sketch of our Presidents from Washington to Garfield, with Official Reports and a Store of Miscellaneous Matter, alike valuable to the Student, Merchant, Mechanic and Farmer.

By ALBERT B. HUSSONG, A. M.

1881.

Entered according to Act of Congress, in the year 1881,
By A. B. HUSSONG,
In the office of the Librarian of Congress, at Washington, D. C.

BLOMGREN, BROS. & CO.,
ELECTROTYPERS,
162 & 164 Clark Street, Chicago.

OTTAWAY & COMPANY,
PRINTERS,
54 & 56 Franklin Street, Chicago.

PREFACE.

It has been our aim to prepare a work of such general information, as we hope will make it a welcome visitor in every household. We present it to the reader in a concise yet clear manner, having spared no pains to be accurate, that the book may be both instructive and reliable. A. B. H.

CONTENTS.

	PAGE.
America	7
Alphabet, Telegraph	142
Authors, Table of	116
Authors, Sketch of	129
Army, Pay of the	100
Advertisements	208-9-10
Cities, Population of	111
Colleges, Founding of	140
Cabinets	14
Conkling, The name of	49
Congressional Vote of Indiana in 1880	67
Congressional Vote of Illinois in 1880	68
Coins, Value of	96
Deadlock, The	48
Drama, The	133
Election, of President	12
Election, of States	77
Exchange	103
Electoral Commission of 1876	40
Garfield, Assassination of	199
Governors, Terms of Office and Salaries	77
Grain and Produce, Table of	109
House, Offices of the	76
Interest, Rate of	70
Illinois, Representatives of	75
Indiana, Representatives of	76
Independence, Declaration of	147
Inventors, American	112
Language, Origin of	123
Language, English	125
Legislatures, Meeting of	70
Mountains, Peaks of	104
Presidents, Sketches of	14
Patent Fees	110

CONTENTS.

	PAGE.
Population of the United States by Races in 1880	60
Population of the Territories in 1870 and 1880	94
Population of Ohio, Indiana, Illinois, Missouri and Kansas by Counties in 1870-80	61
Popular and Electoral Vote of all the Presidents	50
Popular Vote of 1876 and 1880	53
Presidential Vote of Ohio, Indiana, Illinois, Missouri and Kansas by Counties in 1880	54
Rivers, Length of	105
Representatives, House of	75
Rulers, Our	12
Race, Human	121
Salaries	51
Senators Elect to the 47th Congress, 1881-83	74
Senate, Officers of the	74
United States Money	95
United States, Emigration to	99
United States, Supreme Court of	71
United States, Constitution of the	153
United States, Postal Law	143
Universities and Colleges	139
Wood, Weights of	104
World, Religious Divisions of the	109
Washington's Farewell Address	177

AMERICA.

America had been a field of speculation to the inhabitants of the old world from its first discovery. Favorite navigators were equipped and sent forth to explore the NEW WORLD. From the sunny land of Spain came Columbus, Magellan, De Soto and others. From England came the Cabots, Gosnold, Drake, Sir Walter Raleigh and Gilbert. From France sailed out Cartier, Ribault and Francis of La Rogue. Each country was anxious to obtain a foothold upon the soil of the GREAT WEST. James I. was upon the throne of England, Henry IV. was ruling France, and Philip III. was King of Spain when the first English settlement was made at Jamestown, in the year 1607. Charters were granted to companies of these different nations, conveying hundreds of miles of land, entangling boundary lines with each other and thus sowing the seed of discord for future generations to reap with the sword. The story of our land had been told and emigrants poured in upon our shores from the various nations of Europe. The earliest settlements affected by the English colonies were those of

Connecticut,	at Windsor,	in the year	1633.
Delaware,	at Christiana,	" "	1638.
Georgia,	at Savannah,	" "	1733.
Maryland,	at St Mary's,	" "	1634.
Massachusetts,	at Plymouth,	" "	1620.
New York,	at New York,	" "	1614.
New Hampshire,	at Portsmouth,	" "	1623.

New Jersey,   at Elizaoethtown, in the year 1665.
North Carolina, at Albemarle Sound, "   " 1663.
Pennsylvania,   at Philadelphia,   "   " 1683.
Rhode Island,   at Providence,   "   " 1636.
South Carolina, at Charleston,   "   " 1670.
Virginia,   at Jamestown,   "   " 1607.

From the settlement of Jamestown until the beginning of the Revolutionary War—covering a period of one hundred and sixty-seven years—America had been a land of contention; her colonies were at war with the hostile Indian at home, and involved in nearly all the wars between the old countries in Europe; her commerce was restricted to such channels as pleased the will of a king; many of the subordinate rulers assumed the power and arrogance of a monarch, and manifested but little care for the toil and hardship of the laboring masses.

England had been at war with France from 1689 to 1697, known as KING WILLIAM'S WAR. Again, in 1702, declared war against France and Spain, known in Europe as the WAR OF THE SPANISH SUCCESSION, in America as QUEEN ANNE'S WAR. A treaty closed hostilities in 1713. England was again at war with Spain in 1739 until 1744, when she again clashed arms with France, known in Europe as the WAR OF THE AUSTRIAN SUCCESSION, in America as KING GEORGE'S WAR. This was concluded by a treaty of peace in 1748. From 1754 to 1763 occurred the FRENCH AND INDIAN WAR. This was by far the most severe and destructive of all preceding wars in which the colonies were involved. In 1760 George III. ascended the throne of England. It was under his rigid and tyrannical rule that the English Parliament declared it their intention to tax the American colonies to defray, in a great part, the expenses of the French and Indian war—a war into which England had forced her colonies—not so much by reason of disputed claims as her avenging spite to France. One obnoxious measure followed another, until the English colonies became convinced that the only way to put an end to the submission was to break off all allegiance from the mother country.

Now, the colonies, who but a few years before had aided the English in the prosecution of a war against France, were determined to free themselves from the oppressive yoke of Great Britain.

The thirteen states, lacking Georgia, assembled at Philadelphia, September 5, 1774, and drafted a "Declaration of Colonial Rights." This only stirred the ire of England to measures more severe, until as a last resort, Congress, July 4, 1776, at the risk of their lives, broken-up homes and desolated fields that would follow from the wrath of an offended king, drew up and signed the "Declaration of Independence," which concludes, "With a firm reliance on the protection of DIVINE PROVIDENCE, we mutually pledge to each other our LIVES, our FORTUNES, and our SACRED HONOR."

The first battle had already been fought at Lexington, Massachusetts, April 19, 1775, and the war had been waging more than two years when the Articles of Confederation were submitted to the State Legislatures for ratification or rejection; they were returned, requesting some changes, to which Congress readily acceded, when they were adopted and signed July 9, 1778, and became the law of the new government. This war, which lasted from 1774 to 1781, is known as the *Revolutionary*, or the *War of American Independence*.

In May, 1787, a Constitutional Convention assembled at Philadelphia and from a revision of the "Articles of Confederation" framed the "Constitution of the United States," which was ratified by the thirteen original States in the following order:

Delaware, December 7, 1787.
Pennsylvania, December 12, 1787.
New Jersey, December 18, 1787.
Georgia, January 2, 1788.
Connecticut, January 9, 1788.
Massachusetts, February 6, 1788.
Maryland, April 28, 1788.
South Carolina, May 3. 1788.
New Hampshire, June 21, 1788.
Virginia, June 26, 1788.
New York, July 26, 1788.
North Carolina, November 21, 1788.
Rhode Island, May 29, 1790.

One hundred and four years have passed away since the adoption of our constitution. What a change has taken place! The development of our institutions stands without a parallel; the advance and rapid growth of our country throws a shadow over the civilized nations of Europe. Then, with but thirteen states and three million of people; now, we have thirty-eight states and fifty million of people; then, the planters of the South cleaned their cotton by hand, one pound being a day's work for a man; now, a single machine, the cotton-gin, performs the labor of five thousand persons. Then, the great wheat-fields were harvested with a reaping-hook, an acre per day to a man; now, a boy cuts and binds ten acres per day with the self-binding reaper. Then, telegraphy was unknown. France could declare a war, wage and almost end it before we would hear of it; now, we can read every morning at our breakfast-tables that which occurred in Europe the day before. Then, it required from one to five months to sail across the Atlantic; now, a steamer makes the trip from New York to Queenstown, 3,000 miles, in seven days. Then, not a mile of railway; now, we have 100,000 miles; besides the internal improvements, our vacant lands are becoming possessed. The total emigration to the United States for sixty years back numbers 10,000,000. For 1879 it numbered 175,000; for 1880, 500,000; and from the present indications, one million of foreigners will be landed upon our shores for the year 1881. Sixty thousand emigrants landed at New York during the month of April, 1881. In Germany emigration has almost assumed the character and magnitude of an exodus. Vessels are engaged weeks before their time of departure, and sister countries will doubtless follow suit. What else does this argue than that emperors, kings, and princes are becoming intolerable; that taxation and starvation to do them honor is becoming unpopular among the toiling millions; that Queen Victoria with her idle wealth of $15,000,000 of royal gold and silver services at Windsor Castle *alone*, while poverty reigns in other parts of the Island, only rules badly; the same may be said of other crowned heads of Europe. No wonder they come!

Our country hails the *industrious* of all nations— those who wish to engage in the various pursuits of American industry, to acquire wealth, educate their children and become a happy and prosperous people in the land of their adoption.

Our commerce crosses every sea and enters the different ports of the world; our inland traffic upon the silver lakes, bays and rivers is immense; our resources of wealth are inexhaustible; our broad prairies and fertile valleys are opulent, and our mines abound in riches. With an honest administration from the "government of the people, by the people and for the people," our land bids fair to become the most powerful nation of all Christendom.

OUR RULERS.

THE PRESIDENT, SUPREME COURT AND CONGRESS.

In these three branches we have the Executive, Judicial, and Legislative bodies of our Government, from which eminate the laws that govern the affairs of our nation. By an act of Congress, March 1, 1792, and as amended in 1845, fixes a uniform day of election for Electors of President and Vice-President—Tuesday after the first Monday in November preceding the expiration of a presidential term. These Electors meet at the capital of their respective states on the first Wednesday of December following, "and vote by ballot for President and Vice-President; naming in their ballots the person voted for as President, and in distinct ballots the person voted for as Vice-President; they shall make distinct lists of all persons voted for as President and of all persons voted for as Vice-President, and of the number of votes for each, which lists they shall sign and certify, and transmit, sealed, to the seat of the Government of the United States, addressed to the President of the Senate." On the second Wednesday of February following, the President of the Senate shall open the certificates in the presence of both Houses assembled, and the votes shall then be counted; the persons voted for as President and Vice-President, having a majority of the whole number of electors appointed, shall be declared elected. The newly-elected officers receive the oath of office on the 4th of March following.

Upon taking his seat, the President of the United States appoints the members of his cabinet, which are sent in to the Senate for confirmation. At present they number seven, to wit: a Secretary of State, Secretary of Treasury, Secretary of War, a Secretary of the Navy, Secretary of the Interior, Postmaster-General and the Attorney-General. The Secretary of the Navy was not added until 1798. The Postmaster-General was not a member of the cabinet until 1829, although some one has acted in that capacity from the first cabinet; and the Secretary of the Interior was not added until 1849. The selection of a cabinet is a matter of much responsibility; more so, even, than a majority of the people imagine. It is no difficult task to find seven men who are *perfectly* willing to accept these positions. The President is not put to a horse-back ride of thirty miles, through mud and snow, to call on a man and induce him to come forward. But to choose these members best fitted for the different departments of the government—men whose honesty, uprightness and integrity is beyond all question—requires a man of no mean ability; a man with a clear intellect; in short, a statesman. These members should be selected with great care, being chosen by the President as so many advisers, whose wise counsel is intended to assist him in executing the laws of the land.

Now days, the gifts within the hands of the President have become a burden to him. He is beseiged from early morn till late at eve; from the first faint dawn of election till the last fort is taken; and, no doubt, yields often to the powerful influence of friends against the dictates of his better judgment. Certainly, a brilliant and wise administration is desired upon the part of every President, not only as a matter of pride and fact in history, but as a credit abroad, that our nation—abundant in wealth, glowing with intelligence, happy and prosperous—may stand a peer with any civilized nation in the world.

WASHINGTON'S CABINET.
1789–1797.

Thomas Jefferson,	Va.,	Secretary of State,
Edmund Randolph,	Va.,	Secretary of State.
Timothy Pickering,	Mass.,	Secretary of State.
Alexander Hamilton,	N. Y.,	Secretary of Treasury.
Oliver Wolcott,	Ct.,	Secretary of Treasury.
Timothy Pickering,	Mass.,	Secretary of War.
James McHenry,	Md.,	Secretary of War.
Henry Knox,	Mass.,	Secretary of War.
Samuel Osgood,	Mass.,	Postmaster-General.
Timothy Pickering,	Mass.,	Postmaster-General.
Joseph Habersham,	Mass.,	Postmaster-General.
Edward Randolph	Va.,	Attorney-General.
Wm. Bradford,	Pa.,	Attorney-General.
Charles Lee,	Va.,	Attorney-General.

George Washington was twice President, serving from 1789 to 1797, with John Adams, of Massachusetts, as Vice-President both terms.

Washington was born in Virginia, Westmoreland County, February 22, 1732. His early education was very meager, only such as could be acquired at a common country school; but endowed with mental superiority, and possessed of a practical turn of mind, he soon gained distinction throughout the limits of his native state; his even manner—unswerving in his convictions of right and justice—honest and courteous in dealing with his fellow-man—laid the corner-stone of his future greatness. He held the office of public surveyor when quite a youth, and was only nineteen years of age when made a military inspector of a portion of Virginia. A dispute concerning the territory of the Ohio Valley was now waxing warm between France and Great Britain, the latter having settled principally along the coast, while the French had explored the interior—up the rivers and northwest to the lakes. Before an open declaration of war between the contestants, Governor Dinwiddie sent a message to General St. Pierre, near the Presque Isle on the shore of Lake Erie, in hopes of affecting a peaceful settlement over the disputed claims; but to no purpose. Nothing short of a full surrender of the territory in question to the French would do General St. Pierre. War was inevitable. This message was borne by Washington; this was in the autumn of 1753, and Washington had just passed his twenty-first birthday. None but the bold, patriotic and most daring would have undertaken such a perilous adventure; his journey, which occupied days, weeks, and even months—traveling hundreds of miles through a wild and desolate country, was finally accomplished; and on his safe return, Old Virginia struck glad hands with this worthy and future President of the United States. What then was a most hazardous undertaking—health and life exposed and imperiled—would to-day be a journey of comfort and pleasure; the forest and wilderness shorn of original density, now teem with corn and golden grain; innumerable channels of trade and commerce are plied by the master-hand of Genius; and the iron-horse with its palace cars of art and beauty, bridges the rivers, tunnels the mountains, bounds o'er the prairies, and glides through the valleys from the Atlantic to the Pacific. Colonel Washington was helping suppress this, the French and Indian War, embracing a period of nine years from 1754 to 1763. Great Britain sent over General Braddock to assist the Colonies, with particular instructions to honor provincial colonels as " high privates in the rear rank." Washington took offense at this, resigned and went home. In January, 1756, was married to one Mrs. Martha Custiss; but his services were indispensable and he was prevailed upon to return and assume his duties as colonel. Marching on to Fort du Quesne, Washington fell in again with this high-toned Braddock, whom he advised to go slow, so narrow, rough and hilly was the passage; but Braddock, more brave than judicious; pushed forward his troops into the very jaws of death—an Indian snare, and was handled without gloves; *then* it was, in the lull of battle, after the flower of the English army had been slain, that the fatally-wounded Braddock raised and recognized Colonel Washington, whose coat, too, the enemy had riddled with bullets, but whose life was spared to witness the sad result of Braddock's hasty action. Done with this war, in which the English were finally victorious, Washington was next seen in public life as a member of the first Continental Congress, which met in Philadelphia September 5, 1774.

Ten years had scarcely elapsed, when the Colonies, taxed and sore from the late conflict with the French and Indians, were now upon the verge of a war with Great Britain. Congress reassembled May 10, 1775, during which session Washington was unanimously elected Commander-in-chief of the army; in accepting the position he declined any pay for his services save his necessary expenses, which Congress might defray if it saw fit so to do. Commissioned in June, he took the field in July, 1775; for the grim demons of King George, with their iron fangs and bristling bayonets, were already abroad in the land. For seven long years of cloud and sunshine, snow and storm, success and defeat, this brave man led his army on, until the last foe from the British Isle, Cornwallis, surrendered his entire command; this virtually closed the Revolutionary War. In the winter of 1783 a final *Treaty* was signed at Paris, by ambassadors from England, France, Spain, Holland and the United States. The substance of this treaty is thus briefly condensed by Professor J. C. Ridpath, in his History of the United States:

"A full and complete recognition of the independence of the United States; the recession by Great Britain of Florida to Spain; the surrender of all the remaining territory east of the Mississippi and south of the great lakes to the United States; the free navigation of the Mississippi and the lakes by American vessels; and the retention by Great Britain of Canada and Nova Scotia, with the exclusive control of the St. Lawrence." In December, 1783, Washington resigned and retired to his home at Mount Vernon to rest, as desired, but only to be besieged by friends and letters relative to the formation of a new government. Next we find Washington Chairman of the convention held at Philadelphia in May, 1787, for the purpose of revising the Articles of Confederation, which, by various changes, became our present Constitution of the United States. Washington, having been unanimously elected President, was inaugurated April 30, 1789, at the City Hall of New York; in the fall of 1792 was re-elected, and at the close of his second term was solicited to accept a third, but absolutely declined; resolving to spend the rest of his days in quiet and peace, since "The Rubicon had been crossed" and the foundation for the erection of a mighty republic thoroughly established. His "Farewell Address" was given to the people September 17, 1796; he retired March 4, 1797, to his home at Mount Vernon, where he died on the 14th of December, 1799. The capital of the country remained at New York until 1790; at Philadelphia until 1800, when by an act of Congress, previously passed, a tract of land ten miles square— ceded by Virginia and Maryland—united to form the *District of Columbia*, whereon was built the capital city of the nation. This tract of ten miles square is equal to one hundred square miles; in 1846, thirty-six square miles, on the Virginia side of the Potomac River, was receded to Virginia, thus reducing the *District* to its present area of sixty-four square miles.

JOHN ADAMS' CABINET.

1797-1801.

Timothy Pickering,	Mass.,	Secretary of State.
John Marshall,	Va.,	Secretary of State.
Oliver Wolcott,	Ct.,	Secretary of Treasury.
Samuel Dexter,	Mass.,	Secretary of Treasury.
James McHenry,	Md.,	Secretary of War.
Roger Griswold,	Mass.,	Secretary of War.
George Cabot,	Mass.,	Secretary of Navy.
Benjamin Stoddert,	Md.,	Secretary of Navy.
Joseph Habersham,	Mass.,	Postmaster-General.
Charles Lee,	Va.,	Attorney General.
Theophilus Parsons,	Mass.,	Attorney-General.

John Adams served four years; from 1797 to 1801, with Thomas Jefferson as Vice-President. Adams was born at Quincy, Massachusetts, October 19, 1735; was educated at Harvard, graduating at the age of twenty;

married a Miss Abigal Smith in 1764; was a member of the first Colonial Congress; was three years a minister to England, and Vice-President under Washington. After devoting forty years of service to his country, retired from public life to spend the rest of his days on his farm, where he died on the 4th of July, 1826, at the ripe old age of ninety-one years.

JEFFERSON'S CABINET.

1801–1809.

James Madison,	Va.,	Secretary of State.
Samuel Dexter,	Mass.,	Secretary of Treasury.
Albert Gallatin,	Pa.,	Secretary of Treasury.
Henry Derbon	Mass.,	Secretary of War
Benjamin Stoddert,	Md.,	Secretary of Navy.
Robert Smith,	Md.,	Secretary of Navy.
J. Crowingshield,	Mass.,	Secretary of Navy.
Joseph Habersham,	Mass.,	Postmaster-General.
Gideon Granger,	Ct.,	Postmaster-General.
Levi Lincoln,	Mass.,	Attorney General.
Robert Smith	Md.,	Attorney-General.
John Breckinridge,	Ky.,	Attorney-General.
C. A. Rodney,	Del.,	Attorney-General.

Thomas Jefferson served from 1801 to 1809, with Burr, of New York, and Clinton, of New York, as Vice-Presidents respectively. Jefferson was born in Albermarle County, Virginia, April 2, 1743; was thoroughly educated and entered the profession of law; was married at the age of twenty-nine to Mrs. Martha Skelton; represented his state in the Legislature; was a member of the Continental Congress in 1775, and the father of the Declaration of Independence; was Governor of Virginia in 1779; was five years a Minister to Europe; was the first Secretary of State and the third President of the United States. He and Aaron Burr received the same number of electoral votes, so the election went to the

House, and after a six days contest, Jefferson was chosen on the 36th ballot, and Burr receiving the next highest vote became Vice-President. Jefferson was the first President to serve in the new Capitol at Washington; he was re-elected in 1804; according to the twelfth amendment, the electors were now required for the first time to vote separately for President and Vice-President. He devoted the seventeen years of retired life to study of literature, and died on the 4th of July, 1826.

MADISON'S CABINET.
1809-1817.

Robert Smith,	Md.,	Secretary of State.
James Monroe,	Va.,	Secretary of State.
Albert Gallatin,	Pa.,	Secretary of Treasury.
George Campbell,	Tenn.,	Secretary of Treasury.
Alexander Dallas,	Pa.,	Secretary of Treasury.
William Eustis,	Mass.,	Secretary of War.
John Armstrong,	N. Y.,	Secretary of War.
James Monroe,	Va,	Secretary of War.
W. H. Crawford,	Ga.,	Secretary of War.
Paul Hamilton,	S. C.,	Secretary of Navy.
William Jones,	Pa.,	Secretary of Navy.
B. W. Crowingshield,	Mass.,	Secretary of Navy.
Gideon Granger,	Ct.,	Postmaster-General.
Reuben J. Meigs,	Ohio,	Postmaster-General.
E. A. Rodney,	Del.,	Attorney-General.
William Pinkney,	Md.,	Attorney-General.
Richard Rush,	Pa.,	Attorney-General.

Madison served eight years, from 1809 to 1817, with George Clinton, of New York, the first term, and Elbridge Gerry, of Massachusetts, the second term, as Vice-Presidents. James Madison was born in Virginia, March 5, 1751. Officiated in the Continental Congress; married a Mrs. Todd; was a member of the convention that met in Philadelphia in 1787 to frame a Constitution, and succeeded to the Presidency in time to take charge of a war that was now threatening our country. Great

Britain still held a grudge against the United States, and gave vent to her insolence by violating an international law. Her cruisers were authorized to halt American vessels upon the high seas, go aboard, take all Englishmen found and press them into the English service, without "right or writ." This and many other grievances were bearing down on the American people at the beginning of Madison's administration, which resulted in a war at sea—declared against Great Britain by the United States, on the 19th of June, 1812, and known as the war of 1812. It was in this war that Commodore Oliver Perry, of Rhode Island, at the age of twenty-eight, with nine ships and fifty-four guns, met his antagonist, the British Commodore Barclay, with his six large vessels and sixty-three guns, in open combat on the waters of Lake Erie, September 10, 1813. The line of battle was drawn; both commanders were slow but sure of victory; cool and deliberately Perry threw a "light pounder" at the British for a "fielder;" and it was sufficient. With all their power and might the British navy responded, and shivered Perry's flag-ship almost to atoms; he only escaping in an open boat to another of his ships. Not in the least despondent, however, he drove his fleet into the very midst of the enemy, and poured out volleys of thunder and death, until the British, wrecked and ruined, surrendered. It was in this engagement that originated the phrase, which Perry sent to General Harrison: "We have met the enemy and they are ours." It was in this war that the eternal Jackson fought the famous battle of New Orleans, and that, too, fifteen days after a treaty of peace had been signed between the United States and Great Britain. Jackson had not "heard the news."

Madison retired to private life March 4, 1817, and died at Montpelier, Va., June 28, 1836.

MONROE'S CABINET.
1817–1825.

John Q. Adams,	Mass.,	Secretary of State.
W. H. Crawford,	Ga.,	Secretary of Treasury.
Isaac Shelly,	Ky.,	Secretary of War.
John C. Calhoun,	S. C.,	Secretary of War.
B. W. Crowingshield,	Mass.,	Secretary of Navy.
Smith Thompson,	N. Y.,	Secretary of Navy.
John Rogers,	Mass.,	Secretary of Navy.
Reuben J. Meigs,	Ohio,	Postmaster-General.
John McLean,	Ohio,	Postmaster-General.
Richard Rush,	Pa.,	Attorney-General.
William Wirt,	Md.,	Attorney-General.

James Monroe served eight years, from 1817 to 1825, with Daniel D. Tompkins as Vice-President both terms. Monroe was born in Westmoreland County, Virginia, April 28, 1758; received a partial education at the William and Mary College; took a leading part in the Revolutionary War, and by its close had reached the rank of colonel. In 1782 represented Virginia in the Legislature; applied himself diligently to the study and practice of law, and at the age of twenty-five was elected to Congress; six years later to the United States Senate; was a Minister to France; two terms Governor of Virginia, and Secretary of State under Madison. During his Presidential term peace and quiet prevailed; several states were admitted into the Union, among them Missouri, previous to whose admission a lively discussion arose in Congress. Question: "Should Missouri be admitted as a free state?" In the heat of a long and fierce debate, Henry Clay came to the rescue with his "Missouri Compromise Bill," which put a quietus on slavery for some time. The negative, however, gained the day so far as Missouri was concerned. Monroe died in New York, July 4, 1831.

JOHN QUINCY ADAMS' CABINET.
1825–1829.

Henry Clay,	Ky.,	Secretary of State.
Richard Rush,	Pa.,	Secretary of Treasury.
James Barbour,	Va.,	Secretary of War.
Peter B. Porter,	N. Y.,	Secretary of War.
Samuel L. Southard,	N. J.,	Secretary of Navy.
John McLean,	Ohio,	Postmaster-General.
William Wirt,	Md.,	Attorney-General.

Adams served four years, from 1825 to 1829, with John C. Calhoun, of South Carolina, as Vice-President. He was a son of John Adams, the only instance where a father and son have been chief magistrates of the country. We might predict the same for Robert Lincoln, son of Abraham. *John Quincy Adams* was born at Quincy, Mass., July 11, 1767; graduated at Harvard College at the age of twenty; became an efficient lawyer at the bar of Boston; received many honors of distinction at the hands of Washington, as well as from his own father; represented the United States at different times as Minister to Holland, Berlin and St. Petersburg; was Secretary of State under Monroe. In the contest of 1824 were Jackson, Adams, Clay and Crawford, neither of whom received a majority of the electoral votes. The House of Representatives chose Adams, who aspired for re-election, but his strenuous advocacy of high protective tariff and noted hatred for slavery made his defeat an absolute certainty. He was, however, returned a member to Congress, and died at Washington during its session, February, 12, 1848, at the age of eighty-one years.

ANDREW JACKSON'S CABINET.
1829-1837.

Martin Van Buren,	N. Y.,	Secretary of State.
Edward Livingston,	La.,	Secretary of State.
Louis McLane,	Del.,	Secretary of State.
John Forsythe,	Ga.,	Secretary of State.
Samuel D. Ingham,	Pa.,	Secretary of Treasury.
Louis McLane,	Del.,	Secretary of Treasury.
William J. Duane,	Pa.,	Secretary of Treasury.
Roger B. Taney,	Md.,	Secretary of Treasury.
Levi Woodbury,	N. H.,	Secretary of Treasury.
John H. Eaton,	Tenn.,	Secretary of War.
Lewis Cass,	Mich.,	Secretary of War.
John Branch,	N. C.,	Secretary of Navy.
Levi Woodbury,	N. H.,	Secretary of Navy.
Mahlon Dickerson,	N. J.,	Secretary of Navy.
William F. Barry,	Ky.,	Postmaster-General.
Amos Kendall,	Ky.,	Postmaster-General.
John M. Berrien,	Ga.,	Attorney-General.
Roger B. Taney,	Md.,	Attorney-General.
Benjamin F. Butler,	N. Y.,	Attorney-General.

Andrew Jackson served from 1829 to 1837, with John C. Calhoun as Vice-President the first term and Martin Van Buren the second term. Jackson was born in North Carolina, March 15, 1767. Left an orphan boy, his education was made up of piece-meals here and there. He first tried the harness trade; then undertook to amuse the children of a "deestrict skool;" afterwards moved to Tennessee and located at or near Nashville and began the practice of law. At the age of twenty-four he married Mrs. Rachel Robarbs. He soon reached the legislative halls of Congress and then the United States Senate. He was one of the coolest and most determinate men of his time; could settle a difficulty with an enemy by a friendly glass or a duel, and did not care a d—n which. As has been noticed, he took a hand in the war of 1812, in which he won his fame as a military genius. In 1817 the Seminole Indians of Georgia began to war on the settlers, when Jackson passed down with a few hundred Tennessee riflemen, and they soon said "enough." Still pursuing his course down into Florida he seized, court-marshaled and executed two British culprits, whom, it seemed, had been main leaders in attempting to incite the Indians to an insurrection; and this little incident came very near plunging America into a war with Great Britain. Many of Jackson's enemies—more eager for fame than bold—sprung the question of Jackson's right of procedure on such a provocation; but on went Jackson until he had fleeced the Indians and drove the Spaniards to an island of the sea. Jackson was a candidate for the Presidency in 1824 but was defeated; in 1828 he was elected. Now his time had come. Opponents to his method of doing business in former years were numerous and had made his nomination a subject of jest and ridicule; but as soon as he had taken the oath of office he began his slaughter, turning men out high and low, until six hundred and eighty-three "heads had tumbled into the basket." "*I'll show you, by the* ETERNAL, *who's running this government.*"

He was re-elected in 1832. His heroic deeds were known far and wide; his vindictiveness was marked by the least casual observer; his measures were executed with an iron will, and his enemies were made to fear rather than trifle with him. At the expiration of his term he retired to his farm, and died on the 8th of June, 1845.

MARTIN VAN BUREN'S CABINET.
1837–1841.

John Forsythe,	Ga.,	Secretary of State.
Levi Woodbury,	N. H.,	Secretary of Treasury.
Joel R. Poinsett,	S. C.,	Secretary of War.
Mahlon Dickerson,	N. J.,	Secretary of Navy.
James K. Paulding,	N. Y.,	Secretary of Navy.

Amos Kendall,	Ky.,	Postmaster-General.
John M. Niles,	Ct.,	Postmaster-General.
Benjamin F. Butler,	N. Y.,	Attorney-General.
Felix Grundy,	Tenn.,	Attorney-General.
H. D. Gilpin,	Pa.,	Attorney-General.

Martin Van Buren served from 1837 to 1841, with R. M. Johnson, of Kentucky, as Vice-President. He was born in New York, December 5, 1782. Unlike his predecessor, he enjoyed the advantages of a good education, and entered the profession of law with flattering prospects before him; soon attracted attention at the bar, represented his state in the Legislature; as Attorney-General, Governor, twice United States Senator, and on the 4th of March, 1837, inaugurated President of the United States. Was twice a candidate for re-election and both times defeated, when he withdrew from public life and died on his farm, July 24, 1862.

WILLIAM H. HARRISON'S CABINET.

ONE MONTH.

Daniel Webster,	Mass.,	Secretary of State.
Thomas Ewing,	Ohio,	Secretary of Treasury.
John Bell,	Tenn.,	Secretary of War.
George E. Badger,	N. C.,	Secretary of Navy.
Francis Granger,	N. Y.,	Postmaster-General.
John J. Crittenden,	Ky.,	Attorney-General.

Harrison served from March 4, 1841 to April 4, 1841, with John Tyler, of Virginia, as Vice-President. Harrison was born in Virginia February 9, 1773; was educated at Hampden, Sidney College. He chose to win his laurels in defending the early settlers of the great northwestern territory against the savage red man; he was a plain, economical, pure-hearted man, and soon won the esteem of his commanders and the good will of his fellow countrymen. He speedily rose to distinction and was made Governor of the "Indian Territory," with the old French town of Vincennes as the capital of this vast area. After several decisive battles, among which was that of Tippecanoe, Harrison, now Commander-in-chief, taught the thirsty, scalping villains how to submit by the force of arms. He was afterwards sent to the United States Senate from Ohio; was a candidate for President in 1836, but defeated; again the Whigs took him up in 1840 and elected him with 234 electoral votes against 60; died in just one month after his inauguration.

JOHN TYLER'S CABINET.

1841—1845.

Daniel Webster,	Mass.,	Secretary of State.
Hugh S. Legare,	S. C.,	Secretary of State.
Abel P. Upshur,	Va.,	Secretary of State.
John C. Calhoun,	S. C.,	Secretary of State.
Thomas Ewing,	Ohio,	Secretary of Treasury.
Walter Forward,	Pa.,	Secretary of Treasury.
John C. Spencer,	N. Y.,	Secretary of Treasury.
George M. Bibb,	Ky.,	Secretary of Treasury.
John Bell,	Tenn.,	Secretary of War.
John C. Spencer,	N. Y.,	Secretary of War.
James M. Porter,	Pa.,	Secretary of War.
William Wilkins,	Pa.,	Secretary of War.
George E. Badger,	N. C.,	Secretary of Navy.
Abel P. Apshur,	Va.,	Secretary of Navy.
David Henshaw,	Mass.,	Secretary of Navy.
G. W. Gilmer,	Va.,	Secretary of Navy.
John Y. Mason,	Va.,	Secretary of Navy.
John J. Crittenden,	Ky.,	Attorney-General.
Hugh S. Legare,	S. C.,	Attorney-General.
John Nelson,	Md.,	Attorney-General.
Francis G. Granger,	N. Y.,	Postmaster-General.
Charles A. Wickliffe,	Ky.,	Postmaster-General.

After the death of Harrison, Tyler accordingly became President; he served from 1841 to 1845, a full term lacking one month. Tyler was born in Virginia,

March 29, 1790; was educated at the William and Mary College; was five terms a member of the State Legislature; was three terms in Congress; Governor of his native state and a Senator of the United States. At the opening of the Rebellion he cast his lot with the Southern Confederacy, representing it in its Congress; he died at Richmond in January, 1862.

JAMES K. POLK'S CABINET.

1845–1849.

James Buchanan,	Pa.,	Secretary of State.
Robert J. Walker,	Miss.,	Secretary of Treasury.
William M. Marcy,	N. Y.,	Secretary of War.
George Bancroft,	Mass.,	Secretary of Navy.
John Y. Mason,	Va.,	Secretary of Navy.
Cave Johnson,	Tenn.,	Postmaster-General.
John Y. Mason,	Va.,	Attorney-General.
Nathan Clifford,	Me.,	Attorney-General.
Isaac Toucey,	Ct.,	Attorney-General.

James K. Polk served from 1845 to 1849 with George M. Dallas, of Pennsylvania, as Vice-President. He was born in North Carolina on November 2, 1795; was educated at the University of the state; afterwards moved to and settled near Nashville, Tenn. At the age of twenty-nine married Miss Sarah Childress, of Columbia; his course was first Legislator, next Governor of the state; seven terms a member in Congress, and in 1845 inaugurated President of the United States. At the beginning of his administration a war was brooding between the United States and Mexico; the prime *causa belli* grew out of a dispute concerning the boundary line between the two countries; Mexico claimed the territory between the Neuces and the Rio Grande Rivers; while the Americans considered the Rio Grande the dividing line. War was declared in May, 1846; Generals Scott and Taylor soon proved too much for them; into the hands of Scott fell the Mexican army, together with the Capital of Mexico; a treaty of peace followed, and was ratified in March, 1848. The boundary line was fixed to run from the mouth of the Rio Grande to the south line of New Mexico, along this territory west and north to the River Gila, which forms the line to the Colorado, thence west to the Pacific. Mexico relinquished her claims to the disputed territory together with New Mexico and Upper California, for all of which the United States paid Mexico $15,000,000, and her indebtedness to American citizens, amounting to $3,500,000 more. Polk died June 15, 1849.

TAYLOR'S CABINET.
March 4, 1849–July 4, 1850.

J. M. Clayton,	Del.,	Secretary of State.
W. M. Meredith,	Pa.,	Secretary of Treasury.
George W. Crawford,	Ga.,	Secretary of War.
Wm. B. Preston,	Va.,	Secretary of Navy.
Thomas Ewing,	Ohio,	Secretary of Interior.
Jacob Collamer,	Vt.,	Postmaster-General.
Reverdy Johnson,	Md.,	Attorney-General.

Zachary Taylor served one year and four months, with Milliard Fillmore, of New York, as Vice-President. Taylor was born in Virginia, November 2, 1784; received little or no education; moved with his parents to Kentucky and selected a home near the present site of Louisville; preferring a military life to that of farming he entered the army; received a commission at the age of twenty-four, and set out to help General Harrison subdue the Indians in the territory of which he was then Governor. One of his most noted acts of bravery in this war, was the defense of Fort Harrison; left in charge and but twenty-eight years of age, was unexpectedly attacked by an overpowering force of Indians; but he vigorously repulsed the enemy and "held the fort." He took an active part in the Seminole War, and gained several signal victories over the Indians. At the breaking out of the Mexican War he was dispatched to the scene of battle; one of his most brilliant victories was that of Buena Vista, with the odds of four to one against him, completely demoralized and defeated the flower of the Mexican army under the command of Santa Anna. Shortly after the close of this war, Taylor was called to preside over the American people, but only lived a short time; passing away July 4, 1850; died at Washington.

FILLMORE'S CABINET.

July 4, 1850–March 4, 1853.

Daniel Webster,	Mass.,	Secretary of State.
Edward Everett,	Mass.,	Secretary of State.
Thomas Corwin,	Ohio,	Secretary of Treasury.
Charles M. Conrad,	La.,	Secretary of War.
William A. Graham,	N. C.,	Secretary of Navy.
John P. Kennedy,	Md.,	Secretary of Navy.
Alexander H. Stuart,	Va.,	Secretary of Interior.
Nathan A. Hall,	N. Y.,	Postmaster-General.
Samuel D. Hubbard,	Ct.,	Postmaster-General.
John J. Crittenden,	Ky.,	Attorney-General.

By the death of Taylor, Fillmore became President, and served out the balance of the term, from July 4, 1850, to March 4, 1853.

Millard Fillmore was born in New York, January 17, 1800; like Taylor his schooling was limited for the want of means; first apprenticed to a clothier, he employed his odd moments in the study of books; followed teaching a while, and then took up the study of law. He soon built up a large and lucrative business at the bar of Aurora; was a member of the Legislature at the age of twenty; afterwards twice in Congress, and then Vice-President; was a candidate in 1856 for President, nominated by the American or "Know-Nothing party;" he only received the electoral vote of one state—Maryland, while his popular vote was nearly nine hundred thousand; he died at Buffalo, N. Y., March 8, 1874.

PIERCE'S CABINET.

1853–1857.

William L. Marcy,	N. Y.,	Secretary of State.
James Guthrie,	Ky.,	Secretary of Treasury.
Jefferson Davis,	Miss.,	Secretary of War.
J. C. Dobbin,	N. C.,	Secretary of Navy.
Robert McClellan,	Mich.,	Secretary of Interior.
James Campbell,	Pa.,	Postmaster-General.
Caleb Cushing,	Mass.,	Attorney-General.

Franklin Pierce served from 1853 to 1857, with William R. King, of Alabama, as Vice-President; King received the oath of office in Cuba, where he was traveling for his health; returning home he died in Alabama, April 18, 1853, a little more than a month after receiving the oath of office; and the *first* instance of the death of a Vice-President. *Pierce* was born in New Hampshire, November 23, 1804; was educated at Bowdoin College; studied law at Hillsborough; was a member of the Legislature; at twenty-nine elected to Congress; afterwards United States Senator, then President; retired to Concord, where he died October 8, 1869.

BUCHANAN'S CABINET.

1857–1861.

Lewis Cass,	Mich.,	Secretary of State.
Jeremiah S. Black,	Pa.,	Secretary of State.
Howell Cobb,	Ga.,	Secretary of Treasury.
Philip F. Thomas,	N. Y.,	Secretary of Treasury.

John A. Dix,	N. Y.,	Secretary of Treasury.
John B. Floyd,	Va.,	Secretary of War.
Joseph Holt,	Ky.,	Secretary of War.
Isaac Toucey,	Ct.,	Secretary of Navy.
Jacob Thompson,	Miss.,	Secretary of Interior.
Aaron V. Brown,	Tenn.,	Postmaster General.
Joseph Holt,	Ky.,	Postmaster-General.
Horatio King,	Ky.,	Postmaster-General.
Jeremiah S. Black,	Pa.,	Attorney-General.
Edwin M. Stanton,	Pa.,	Attorney-General.

James Buchanan served from 1857 to 1861, with John C. Breckenridge, of Kentucky, as Vice-President. Buchanan was born in Pennsylvania, April 22, 1791; with a fair education he began the study of law; in 1814 was chosen to the Legislature; at the age of twenty-nine was sent to Congress, where he remained for ten years; was twelve years a member of the United States Senate; Secretary of State under Polk; Minister to England under Pierce, and elected President in 1856. For many years the slave question had been a menace to our free institutions; each year added fresh fuel to the aggravated South; the admission of new states kept Congress in a continual broil; the repeal of the "Missouri Compromise Act" four years previous, was still a source of great disorder to the people of Kansas Territory; conventions and legislative assemblies were plentiful and turbid, out of which grew the Lecompton Constitution, nowise approved by the Republicans, but, sanctioned by Congress and the President, became law. It was during Buchanan's administration that John Brown, of Kansas, was executed in Virginia, whither he had repaired to incite the slaves to insurrection; this act created the bitterest feeling in the South towards the North; a war was no longer a question of time, but was already on our hands. Thus stood the state of affairs when Mr. Buchanan retired; he died at Wheatland, Pa., June 1, 1868.

LINCOLN'S CABINET.

1861–1865.

FIRST TERM.

William H. Seward,	N. Y.,	Secretary of State.
Salmon P. Chase,	Ohio,	Secretary of Treasury.
William P. Fessenden,	Me.,	Secretary of Treasury.
Hugh McCulloch,	Ind.,	Secretary of Treasury.
Simon Cameron,	Pa.,	Secretary of War.
Edwin M. Stanton,	Pa,	Secretary of War.
Gideon Wells,	Ct.,	Secretary of Navy.
Caleb B. Smith,	Ind.,	Secretary of Interior.
John P. Usher,	Ind.,	Secretary of Interior.
Montgomery Blair,	Md.,	Postmaster-General.
William Dennison,	Ohio,	Postmaster-General.
Edwards Bates,	Mo.,	Attorney-General.
James Speed,	Ky.,	Attorney-General.

SECOND TERM.

March 4, 1865—April 14, 1865.

William H. Seward,	N. Y.,	Secretary of State.
Hugh McCulloch,	Ind.,	Secretary of Treasury.
Edwin M. Stanton,	Pa.,	Secretary of War.
Ulysses S. Grant,	Ill.,	Secretary of War.
J. M. Scofield,	—	Secretary of War.
Gideon Wells,	Ct.,	Secretary of Navy.
James Harlan,	Ia.,	Secretary of Interior.
Orville H. Browning,	Ill.,	Secretary of Interior.
James Speed,	Ky.,	Attorney-General.
Henry Stanberry,	Ohio,	Attorney-General.
William M. Everett,	N. Y.,	Attorney General.
William Dennison,	Ohio,	Postmaster-General.
Alexander W. Randall,	Wis,	Postmaster-General.

Abraham Lincoln served from 1861 to 1865, with Hannibal Hamlin, of Maine, as Vice-President. He served on his second term from March 4, 1865, to April 14, 1865, with Andrew Johnson, of Tennessee, as Vice-President.

Lincoln was born in Kentucky, February 12, 1809; in 1816 removed with his parents to Spencer County, Indiana, where they resided for some thirteen years,

when they moved to Illinois, in Decatur County.
Abraham was now twenty-one years old, and while he
had been deprived of a thorough education, he had
devoted all his spare moments to a close study of books;
he worked on a farm two years; took a part in the
Black Hawk War; was elected to the Legislature at the
age of twenty-five; began the practice of law at Springfield in 1837; stumped the state in 1844 for Henry
Clay, who was defeated by Mr Polk; three years later
was elected to Congress; in 1848 was defeated for the
the United States Senate, and again ten years later was
defeated by Douglas for the same position; in 1860 was
elected President on a minority vote of 944,000, owing
to a "split" in the Democratic party. The history of
Lincoln's administration, whose lot it was to carry
through one of the most formidable wars of modern
times, is still fresh in the memories of men, and his
course is endorsed by the intelligence of the nation,
regardless of party ties. We find him entering the
scene of conflict in a calm yet determined manner, trying by all honorable means to dissuade the South from
its disloyal course and prevent the terrible calamity of
war; but moral pursuasion was out of the question—
"patience had ceased to be a virtue," and nothing but
a clash of arms would do the sunny South; they got it.
The sentiment of the North and West was largely in
the majority for restoring and maintaining the Union
as it was, by peaceful means if possible; by war if
necessary. Thousands of Democrats, as well as Republicans, shouldered their muskets and marched side by
side upon the field of battle. After four years of struggle the South succumbed and slavery was wiped out
forever from the American soil. Lincoln was re-elected

in the fall of 1864, receiving 212 electoral votes against 21 cast for McClelland. In this election eleven states did not vote, namely: Alabama, Arkansas, Florida, Georgia, Louisiana, Mississippi, North Carolina, South Carolina, Tennessee, Texas and Virginia. These eleven states seceded in the following order:

 South Carolina, December 20, 1860.
 Mississippi, January 9, 1861.
 Florida, January 10, 1861.
 Alabama, January 11, 1861.
 Georgia, January 18, 1861.
 Louisiana, January 26, 1861.
 Texas, February 1, 1861.
 Virginia, April 17, 1861.
 Arkansas, May 6, 1861.
 Tennessee, May 7, 1861.
 North Carolina, May 20, 1861.

At the beginning of the Rebellion the slave-holding states included the above and in addition: Delaware, Maryland, Kentucky and Missouri—fifteen in all.

Lincoln did not live to carry out his policy of another term; was assassinated by J. Wilkes Booth on the night of April 14, 1865, at Ford's Theatre, Washington. Never in the history of our nation were a great and intelligent people stricken with such sorrow. Foreign nations joined in grief with our bereaved people. The world knew him to praise him in life, and bitterly wept for him in death. His noble deeds have found a place in the hearts of nations, that will remain bright on the tablets of memory, and be cherished in all ages to come. He was buried in Oak Ridge Cemetery, Springfield, Ill. By the death of Lincoln, Andrew Johnson became President, and entered upon the duties of office with the newly-appointed cabinet of Lincoln's second term.

Johnson was born in Raleigh, North Carolina, December 29, 1808. He possessed but little of this world's goods—no dishonor but very inconvenient; his education was neglected; learned the tailor trade at an early age; moved to Grenville, Tenn., in 1828; here was married, and by his intelligent wife taught the rudiments of the common school branches—not being able to write before his marriage. His first deal in politics resulted in his election to the council of his native village. From this he gradually rose—a mayor; a Legislator; twice Governor of Tennessee; ten years in Congress; United States Senator; Vice-President in 1864, and by Lincoln's death became President of the United States; a remarkable case in history where a man born in poverty—destitute of learning, without money or friends, rose from an apprenticeship to the highest gift in the hands of the American people. But, however well meant, his administration was very unpopular with both branches of Congress. To bind up the nation's wounds, restore the Southern States to their former status and pay the great war debt, now devolved upon our representatives and our chief magistrate. With all the experience of the past, the necessities of the present, and an outlook for the future of our nation, Congress set to work at its task. But a wide difference of opinion existed between Congress and the President—the former devising a way for the readmission of the seceded states; the latter claiming that the states had never been out of the Union. It soon became evident by a few vetoes that many measures of Congress were not in accordance with the President's views and could not become laws; and a continual disapproval of Congressional bills of reconstruction, and Edwin M. Stanton's dismissal as Secretary of War, lead, finally, to a bill of impeachment, originating in the House but sent to the Senate for trial, presided over by Salmon P. Chase, Chief Justice of the United States. The trial began March 23, 1868, and lasted sixty days. Two-thirds majority was necessary to impeach; it failed by the lack of *one* vote. After the expiration of his Presidential term, Johnson was returned to the United States Senate, and only served a short time when he died, July 30, 1875.

GRANT'S CABINET.
1869–1877.

FIRST TERM.

Elihu B. Washburne,	Ill.,	Secretary of State.
Hamilton Fish,	N. Y.,	Secretary of State.
George S. Boutwell,	Mass.,	Secretary of Treasury.
John A. Rawlins,	Ill.,	Secretary of War.
William W. Belknap	Ia.,	Secretary of War.
Adolph E. Borie,	Pa.,	Secretary of Navy.
George M. Robeson,	N. J.,	Secretary of Navy.
Jacob D. Cox,	Ohio,	Secretary of Interior.
Columbus Delano,	Ohio,	Secretary of Interior.
John A. J. Creswell,	Md.,	Postmaster-General.
E. R. Hoar,	Mass.,	Attorney-General.
Amos T. Akerman,	Ga.,	Attorney-General.
George H. Williams,	Oregon,	Attorney-General.

SECOND TERM.

Hamilton Fish,	N. Y.,	Secretary of State.
W. A. Richardson,	Ill.,	Secretary of Treasury.
B. H. Bristow,	Ky.,	Secretary of Treasury.
Lot M. Morrill,	Maine,	Secretary of Treasury.
W. W. Belknap,	Ia.,	Secretary of War.
Alphonso Taft,	Ohio,	Secretary of War.
James D. Cameron,	Pa.,	Secretary of War.
George M. Robeson,	N. J.,	Secretary of Navy.
Columbus Delano,	Ohio,	Secretary of Interior.
Zachariah Chandler	Mich.,	Secretary of Interior.
J. A. J. Creswell,	Md.,	Postmaster-General.
Marshall Jewell,	Ct.,	Postmaster-General.
James N. Tyner,	Ind.,	Postmaster-General.
George H. Williams,	Oregon,	Attorney-General.
Edwards Pierrepont,	N. Y.,	Attorney-General.
Alphonso Taft,	Ohio,	Attorney-General.

Ulysses S. Grant served from 1869 to 1877, with Schuyler Colfax, of Indiana, as Vice-President first term and Henry Wilson, of Massachusetts, the second term.

Grant was born in Ohio, April 27, 1822; was educated at West Point; was a soldier in the Mexican War, in which he received a captain's commission; returned from this and engaged in the leather business with his father at Galena, Illinois. Here he remained until the beginning of the Rebellion, when he was made colonel of the Twenty-first Illinois Regiment. He rapidly rose to lieutenant-general, and finally was commissioned General of the United States Army, with the pay of $13,500 per annum. He won many brilliant victories during the war, which paved his way to the White House. Was unanimously nominated on the first ballot for the Presidency by the Republican National Convention, convened at Chicago, May 20, 1868; was renominated by acclamation in Convention at Philadelphia, June 5, 1872; was both times elected and served the country eight years. In the canvass of 1872 Grant and Wilson's opponents were Greeley and Brown. Since it is the only instance in our history where a candidate for the Presidency, elected or defeated, died before the Electoral College met, probably it might be of interest to show how the electors voted on the Greeley side, as the popular vote gave Greeley six states, with an aggregate of 66 electoral votes. The electors met as prescribed by law and voted as follows: For Hendricks, of Indiana, 42; for Gratz Brown, of Missouri, 18; for Jenkins, of Georgia, 2; for Davis, of Illinois, 1; not counted, 3. Total, 66.

Henry Wilson, Vice-President under Grant, died November 22, 1875, and after much discussion in the Senate, Thomas W. Ferry, of Michigan, was elected Vice-President, and assumed the duties as President of the Senate. Grant retired on the 4th of March, 1877; shortly afterwards made a trip around the globe, receiving greater honors from the crowned heads of Europe than were ever given before to an American citizen. He returned and was a candidate before the Republican National Convention at Chicago, June 2, 1880; he locked horns with James G. Blaine, of Maine, which resulted in the nomination of James A. Garfield, after one week's contest. General Grant, having been elected president of the Mexican Southern railroad, sailed for Vera Cruz, April 3, 1881.

HAYES' CABINET.
1877-1881.

William M. Evarts,	N. Y.,	Secretary of State.
John Sherman,	Ohio,	Secretary of Treasury.
George W. McCrary,	Iowa,	Secretary of War.
Alexander Ramsey,	Minn.,	Secretary of War.
R. W. Thompson,	Ind.,	Secretary of Navy.
Nathan Goff, Jr.,	W. Va.,	Secretary of Navy.
Carl Schurz,	Mo.,	Secretary of Interior.
David M. Key,	Tenn.,	Postmaster-General.
Horace Maynard,	Tenn.,	Postmaster-General.
Charles Devens,	Mass.,	Attorney-General.

Rutherford B. Hayes served from 1877 to 1881, with William A. Wheeler, of New York, as Vice-President. Hayes was born in Ohio, October 4, 1822; entered Kenyon College in 1838 and graduated in 1842, with the first honors of his class; completed a law course at Harvard University, and took up the practice of law at Fremont; removed to Cincinnati in 1849; entered the War of the Rebellion as major of the Twenty-third Ohio Regiment;

in four months was promoted to lieutenant-colonel; in the fall of 1862 was commissioned colonel; in October, 1864, was promoted to brigadier-general; in the spring of 1865 was made brevet-general; while in the field, was elected to Congress by a handsome majority of the voters of his district. This election occurred in October, 1864, the thirty-ninth Congress to which he was elected did not convene until December, 1865, so Hayes remained in the field until the close of the war. In the Fall of 1867, he was elected Governor of Ohio, and again in 1869; in 1872 was defeated for Congress; in 1875 was for the third time elected Governor of Ohio; in June, 1876, was nominated for President of the United States. This was one of the closest contests in the history of our nation; Samuel J. Tilden, of New York, a man of wealth and ambition, was a strong opponent, receiving two hundred and fifty thousand more votes than Hayes, and but *one* vote less in the Electoral College, after giving Hayes the doubtful states. The states in dispute were Florida, Louisiana, and South Carolina; after a long and bitter discussion it was agreed to leave the settlement of the question to a High Commission of fifteen, composed of five Senators, five Representatives, and five members of the Supreme Court. We give the names and political complexion of this High Court:

SENATORS: Geo. F. Edmunds, R., Vt.
Oliver P. Morton, R., Ind.
F. Frelinghuysen, R., N. J.
A. G. Thurman, D., Ohio.
Thomas F. Bayard, D., Del.

REPRESENTATIVES: Geo. F. Hoar, R., Mass.
James A. Garfield, R., Ohio.
Henry B. Payne, D., Ohio.
Eppa Hunton, D., Va.
Josiah G. Abbott, D., Mass.

ASSOCIATE JUSTICES OF THE SUPREME COURT:
Nathan Clifford, D., Me.
Samuel F. Miller, R., Iowa.
William Strong, R., Pa
Stephen J. Field, D., Cal.
Joseph P. Bradley, R., N. J.

As the summing up of this ELECTORAL COMMISSION showed eight Republicans and seven Democrats, the long discussions were just such wind "throwed away." After several days spent in examining evidence, and sometime in debate, the prize was finally awarded to Mr. Hayes, who was inaugurated on Monday, March 5, 1877. Many not satisfied with the decision, denied the validity of his title; be that as it may, it must be admitted that Hayes is a man of great ability, and no one can question his integrity, while the purity of his public and private life has never *yet* been assailed. His last official act was the veto to a bill, entitled, " An act to facilitate the funding of the national debt." He retired March 4, 1881; on his return home, when within eleven miles of Baltimore, the train collided with two empty engines coupled together sweeping along towards Washington at the rate of thirty-five miles an hour; two persons were killed, but the ex-President and party sustained no injury; he reached Fremont, his home, on the 8th of March, 1881, and was received by a large number of citizens. In reply to a welcome address, Hayes said:

"It strikes me that this is a good place to find an answer to the question which is often heard: 'What is to become of the man? What is he to do? Where is his place, who, having been chief magistrate of the republic, retires at the end of his term to private life?' It seems to me the answer is near at hand and sufficient. Let him, like every other good American citizen, be willing promptly to bear his part in every useful work that will promote the welfare, the happiness and the progress of his family, his town, his state, and his country. With this disposition he will have work enough to do, and that sort of work which yields more individual contentment and gratification than the more conspicuous employments of public life, from which he has retired."

Speaking of the new administration, he said: "Let us give it a hearty and generous support. To those who would not have chosen the new administration, let me say, imitate your gallant and patriotic leader, General Hancock, who, by his presence and cordiality at the inauguration, said to his fellow-citizens, let us give to President Garfield a fair hearing and fair play."

GARFIELD'S CABINET.

James G. Blaine,	Me.,	Secretary of State.
William Windom,	Minn.,	Secretary of Treasury.
Robert T. Lincoln,	Ill.,	Secretary of War.
Wm. H. Hunt,	La.,	Secretary of Navy.
S. B. Kirkwood,	Iowa,	Secretary of Interior.
Thomas L. James,	N. Y.,	Postmaster-General.
Wayne MacVeagh,	Pa.,	Attorney-General.

JAMES A. GARFIELD took his seat March 4, 1881, with Chester A. Arthur, of New York, as Vice-President. Garfield was born in Ohio, November, 19, 1831; his father died when James was but two years old, leaving his widow to rear as best she could her four children, of whom James was the youngest. He followed different occupations during his youth to secure means to aid his mother, and prepare himself for entering a high school. His course at the Hiram Institute enabled him to enter the junior class at William College, where he graduated in 1856, at the age of twenty-five. He immediately took a position as teacher at Hiram College; in 1858 was married to Miss Lucretia Rudolph; in 1859 was elected President of Hiram College. At the opening of the Rebellion, he resigned his professorship and entered the army as colonel of the Forty-second Ohio Regiment; was engaged in several battles of the war. After the battle of Middle Creek, in 1862, he was promoted to brigadier-general, and after the battle of Chickamauga was commissioned a major-general. Concerning the "lay of the field" of this disastrous battle, General Garfield says:

"Rosecrans had crossed the Tennessee, and had successfully manœuvred the enemy out of Chattanooga. The greater work remained—to march his own army into that place, in the face of Bragg's army, heavily re-enforced and greatly outnumbering his own.

"The Rossville road—the road to Chattanooga—was the great prize to be won or lost at Chickamauga. If the enemy failed to gain it, their campaign would be an unmitigated disaster; for the gate-way of the mountains would be irretrievably lost. If our army failed to hold it, not only would our campaign be a failure, but almost inevitable destruction awaited the army itself. The first day's battle (September 19), which lasted far into the night, left us in possession of the road; but all knew that next day would bring the final decision. Late at night, surrounded by his commanders, assembled in the rude cabin known as the Widow Glen House, Rosecrans gave his orders for the coming morning. The substance of his order to Thomas was this: 'Your line lies across the road to Chattanooga. That is the pivot of the battle. Hold it at all hazards;

and I will re-enforce you, if necessary, with the whole army.'

"During the whole night the re-enforcements of the enemy were coming in. Early next morning we were attacked along the whole line. Thomas commanded the left and center of our army. From early morning he withstood the furious and repeated attacks of the enemy, who constantly re-enforced his assaults on our left. About noon our whole right wing was broken, and driven, in hopeless confusion, from the field. Rosecrans was himself swept away in the tide of retreat. The forces of Longstreet, which had broken our right, desisted from the pursuit, and forming in heavy columns, assaulted the right flank of Thomas with unexampled fury. Seeing the approaching danger, he threw back his exposed flank toward the base of the mountains and met the new peril."

The author of Garfield's life—Edmund Kirke—adds:

"It must be borne in mind that the Union army had a line of fully four miles, and was operating in a broken country, half forest and half cotton-field, from no one point of which was it possible to take in the movements of the entire forces. Rosecrans had established his headquarters for the day in the rear of his center and right wing, and on one of the foot-hills of Missionary Ridge. He was here about noon, surrounded by General Garfield, Major McMichael, assistant adjutant-general, Major Bond, his senior aid, and half a dozen orderlies, when Captain Gaw, of Thomas' staff, rode up to his headquarters. The captain had been sent by Thomas with a message to General Negley, and had passed the right center just as Wood opened the fatal gap into which Longstreet streamed, breaking McCook's corps into fragments. Reining his horse to the right he got out of the way of the fugitives. A moment before Rosecrans had caught a distant sight of some scattering troops straggling over the hills, and he called out to Captain Gaw, as he approached, 'What troops are those coming down the hill?'

"'They are a part of McCleve's reserve, General—the right center is broken!'

"In a moment more the hills were swarming with a disordered rabble, and turning to his chief-of-staff, this genuine soldier, who had never before lost a battle, cried out, his anxiety photographed in his face, 'Garfield, what shall be done?'

"Cool, clear-headed and intrepid, this glorious man and wonderful soldier took in, on the instant, the whole extent of the disaster. It did not stun him, but instinctively he turned to some one else to show the way out of the emergency.

"As quietly as if on dress-parade Garfield answers, 'Send an order to General Mitchell (commanding the cavalry) to fall back on Chattanooga; send another to the officer in command at Rossville to withdraw his guards, and let the retreating troops pass; and send Captain Gaw to General Thomas, asking him to take command of all the forces, fall back on Rossville and, with McCook and Crittenden, make a stand there, and hold the enemy in check until you can reorganize the broken divisions.'

"Couriers are quickly dispatched with the several messages, and Captain Gaw has set out, when Rosecrans calls him back, and directs him to show them the shortest route to the Chattanooga Valley Road. They set out, through a trackless forest of cedar-brake and bramble, in the direction of Nickajack Trace. Now and then the commanding general halts, turns his head to listen, and says, 'Thomas is still intact;' then moves on in mournful silence. They come to the Dry Valley Road, and find it crowded with a tangled mass of horses, wagons and soldiers, moving briskly but without the energy of a stampede. As they pass this disorganized mass the general's face reflects the humiliation he feels. Then they leave the highway, and go on again over a rugged, trackless waste. At last they reach the Chattanooga Road, three miles by direct route from Thomas, and four from Chattanooga. Here they halt; and now occurs one of those ludicrous incidents which occasionally break the monotony of the severest battle. The government has sent out a commissioner to look into the state of the army—an honorable Senator, whose reputation fills the country. He approaches them now on horseback, through the open field at the south. His horse is at full speed, and he is hatless, his clothing torn and begrimed with dirt, his hands brier-scratched and bleeding, his hair literally on end, and his face the very image of despair. Rosecrans salutes him; but passing the general without recognition, he rides up to one of the orderlies, and says, 'Sir, have you any tobacco about you ?' The soldier takes out a package wrapped in tin-foil, and the Senator says, hurriedly and emphatically, 'I will give you five dollars for this tobacco, sir.' The orderly declines the money, but tells him to keep the tobacco; and then he turns to the commanding general, and, with a wild and vacant look, says, 'Your army has all gone to h—l, sir! Where is Nickajack Gap? I am bound for Bridgeport.' Then, without waiting for an answer, he turns, puts spurs to his horse and gallops down the road northward. The next heard of him he was seated, canted back in his chair, and with his boots on the top of his desk, in the Senate Chamber at Washington."

After this battle Garfield resigned his commission in the army to accept a seat in the House of Representatives, which he held continually from 1863 until 1879, when he was elected United States Senator to succeed Allen G. Thurman whose time, however, did not expire until March, 1881; but, in the meantime, with a "streak of good luck," was elected President of the United States, consequently he never qualified as a Senator, but resigned his seat in the House of Representatives on the 10th of November, 1880, to take his seat as chief executive of the nation on the 4th of March, 1881.

In this canvass there were three candidates in the field, Garfield, Hancock, and Weaver, brought before the people by the Republican, Democratic, and National Parties, respectively. These men were all three Generals in the Union army, gallant, patriotic and brave, either of whom deserved the honor of President of the United States.

The popular vote given to Garfield was 4,442,950; Hancock, 4,442,035; Weaver, including the full Fusion vote of Maine, 372,743; Neal Dow received 10,305 votes on the Temperance ticket; total, 9,268,033. The Electoral College gave Garfield 214 votes and Hancock 155; Garfield, receiving a majority of all the electoral votes cast, was declared elected, and on the 4th of March, 1881, duly inaugurated President of the United States. Notwithstanding snow and slush, the military display was one of the finest known to the City of Washington; the procession consisted of about fifteen thousand men in line, witnessed by fully 150,000 people. At twelve o'clock Vice-President Wheeler swore in Arthur, the newly-elected Vice-President, this ended the Forty-sixth Congress, of which Wheeler was President. Garfield delivered his inaugural address to a sea of faces, at the conclusion of which he received the oath of office, administered by Chief-Justice Waite.

From this 4th day of March, 1881, the Senate stood Democrats, 38, Republicans, 38, Davis, of Illinois, voting with the Democrats, and Mahone, of Virginia, with the Republicans; being a tie, it devolved upon the Vice-President to give the deciding vote, which was done in reconstructing the committees; but when the Senate proposed to elect new officers, the exercise of this right was questioned, and the election was postponed by enough members refusing to vote to break the quorum; this "dead-lock," as it was called, continued until May 4th—60 days, when the Senate went into an executive session. A war between the President and Conkling was waxing warm; agreeable to Senators Conkling and Platt, the five New York nominations were sent to the Senate for conformation or rejection; in addition to these, the President sent in *one* of his own selection— that of Mr. Robertson as Collector of the port of New York; this *presumed* assumption upon the part of the President, without consulting Roscoe, was not imbued with a proper amount of Senatorial usurp, therefore, he declared himself against the President. This was the Genesis of the war. Committees waited upon Mr. Garfield daily, to persuade him to withdraw the name of Robertson; the President listened long and fervently and finally said, "No! Since the right to exercise my constitutional prerogative free from Senatorial dictation is involved, I'll defend myself if Republican Senators band together in support of Conkling." The five New York nominations were withdrawn, but Robertson's name remained. While this Chickamauga wheel in tactics was an unexpected flank movement upon the enemy, he only "gnashed louder his iron fangs and shook his crest of bristling bayonets." Would that we could record the result of this threatening conflict! The reader knows by this time how the matter terminated. But at this writing we cannot tell. It can only be surmised. "For aught I know the next flash of electric fire along the line may tell us that Conkling, with every fibre quivering with the agony of impotent despair, writhes beneath the conquering heel of Garfield. Ere another moon shall wax and wane the brightest star in the galaxy of Senators may fall from the zenith of his glory never to rise again. Ere the modest violets of spring shall close their beauteous eyes at the approach of this summer sun, the genius of civilization may chant the wailing requiem of the proudest Senator that the Empire State has ever possessed, as she scatters her withered and tear-moistened lilies o'er the political tomb of Conkling."

Byron says:

> 'Tis sweet to win, no matter how, one's laurels,
>   By blood or ink; 'tis sweet to put an end
> To strife; 'tis sometimes sweet to have our quarrels,
>   Particularly with a tiresome friend.

In the table no popular vote is given from 1780 to 1820, the electors during that time being chosen by the Legislatures. The following are the figures:

CANDIDATES.		POPULAR VOTE.	ELECTORAL VOTE.
1789—George Washington	Unanimous		
1796—John Adams	Federal		71
Thomas Jefferson	Democrat		69
1800—Thomas Jefferson*	Democrat		73
Aaron Burr	Democrat		73
John Adams	Federal		65
1804—Thomas Jefferson	Democrat		148
C. C. Pinckney	Federal		28
1808—James Madison	Democrat		122
C. C. Pinckney	Federal		47
1812—James Madison	Democrat		128
DeWitt Clinton	Federal		89
1816—James Monroe	Democrat		183
Rufus King	Federal		34
1820—James Monroe	Democrat		-
1824—John Q. Adams*	Federal	105,321	84
Andrew Jackson	Democrat	155,872	99
W. H. Crawford	Democrat	44,282	41
Henry Clay	Whig	46,587	37
1828—Andrew Jackson	Democrat	647,231	178
John Q. Adams	Federal	509,097	83
1832—Andrew Jackson	Democrat	687,502	219
Henry Clay	Whig	530,189	48
John Floyd	Whig	33,108	11
William Wirt	Whig		7
1836—Martin Van Buren	Democrat	761,549	170
Wm. H. Harrison	Whig	736,656	121
1840—Wm. H. Harrison	Whig	1,275,011	234
Martin Van Buren	Democrat	1,135,761	60
1844—James K. Polk	Democrat	1,337,243	170
Henry Clay	Whig	1,299,062	105
Jas. G. Birney	Free Soil	62,300	
1848—Zachary Taylor	Whig	1,360,099	163
Lewis Cass	Democrat	1,220,544	127
Martin Van Buren	Free Soil	291,263	
1852—Franklin Pierce	Democrat	1,601,474	254
Winfield Scott	Whig	1,386,578	42
John P. Hale	Free Soil	155,185	
1856—James Buchanan	Democrat	1,838,169	174
John C. Fremont	Republican	1,341,264	114
Millard Filmore	American	847,534	8
1860—Abraham Lincoln	Republican	1,866,352	180
Stephen A. Douglas	Democrat	1,375,157	
John C. Breckenridge	Democrat	845,763	123
John Bell	American	589,581	

A STORE OF KNOWLEDGE.

CANDIDATES.		POPULAR VOTE.	ELECTORAL VOTE.
1864—Abraham Lincoln	Republican	2,216,067	213
Geo. B. McClellan	Democrat	1,808,725	21
1868—Ulysses S. Grant	Republican	3,015,071	214
Horatio Seymour	Democrat	2,709,613	80
1872—Ulysses S. Grant	Republican	3,597,070	300
Horace Greeley	Democrat	2,834,076	66
1876—Rutherford B. Hayes	Republican	4,033,950	185
Samuel J. Tilden	Democrat	4,284,757	184
Peter Cooper	Greenback	81,740	
Green Clay Smith	Prohibition	9,552	
1880—James A. Garfield	Republican	4,442,950	214
Winfield S. Hancock	Democrat	4,442,035	155
James B. Weaver	National	307,409	
Neal Dow	Prohibition	10,305	

*Elected by the House of Representatives.

Presidents Washington, Monroe, Jackson, Van Buren, Harrison, Taylor, Pierce, Lincoln, Grant, Hayes and Garfield were soldiers in time of war. Only two of the number were wounded—Monroe and Hayes.

SALARIES.

President of the United States	$50,000.
Vice-President of the United States	8,000.
Chief-Justice of the Supreme Court	10,500.
Associate Justices, each	10,000.
Cabinet Officers, each	8,000.
United States Senators, each	5,000.
Members of Congress, each	5,000.
Speaker of the House of Representatives	8,000.

In addition to his salary, each Senator and Congressman is allowed 20 cents per mile for travel each way during sessions, and $125 per annum for stationery, newspapers, etc. Each state is entitled to two Senators, term six years; each state is divided into Congressional districts, the number depending on the population of the state; each district is entitled to one Congressman; the whole number of Congressmen and the two Senators of any state constitute the number of its electoral vote.

STATISTICS.

POPULAR AND ELECTORAL VOTE FOR PRESIDENT IN 1880.

STATES.	Garfield. Rep.	Hancock. Dem.	Weaver. Gr.	Electoral vote.		
				Garfield	Hancock	Total.
Alabama	56,178	90,687	4,642	....	10	10
Arkansas	41,661	60,489	4,079	....	6	6
California	80,348	80,426	3,392	1	5	6
Colorado	27,450	24,647	1,435	3	....	3
Connecticut	67,073	64,417	868	6	....	6
Delaware	14,150	15,183	120	....	3	3
Florida	23,654	27,964	....	....	4	4
Georgia	52,648	102,522	481	....	11	11
Illinois	318,037	277,321	26,358	21	....	21
Indiana	232,164	225,528	12,986	15	....	15
Iowa	183,904	105,845	32,327	11	....	11
Kansas	121,520	59,789	19,710	5	....	5
Kentucky	104,550	147,999	11,498	....	12	12
Louisiana	31,891	65,310	442	....	8	8
Maine	74,039	65,171	4,408	7	....	7
Maryland	78,515	93,706	818	....	8	8
Massachusetts	165,205	111,960	4,548	13	....	13
Michigan	185,190	131,300	34,795	11	....	11
Minnesota	93,903	53,315	3,267	5	....	5
Mississippi	34,854	75,750	5,797	....	8	8
Missouri	153,567	208,609	35,045	....	15	15
Nebraska	54,979	28,523	3,853	3	....	3
Nevada	8,732	9,611	....	....	3	3
New Hampshire	44,852	40,794	528	5	....	5
New Jersey	120,555	122,565	2,617	....	9	9
New York	555,544	534,511	12,373	35	....	35
North Carolina	115,878	124,204	1,136	....	10	10
Ohio	375,048	340,821	6,456	22	....	22
Oregon	20,619	19,948	249	3	....	3
Pennsylvania	444,704	407,428	20,668	29	....	29
Rhode Island	18,195	10,779	236	4	....	4
South Carolina	58,071	112,312	566	....	7	7
Tennessee	107,677	128,191	5,916	....	12	12
Texas	57,845	156,228	27,405	....	8	8
Vermont	45,090	18,181	1,212	5	....	5
Virginia	84,020	127,976	139	....	11	11
West Virginia	46,243	57,391	9,079	....	5	5
Wisconsin	144,397	114,634	7,980	10	....	10
Total	4,442,950	4,442,035	307,409	214	155	369

STATES	President—1876.			President—1880.		
	Tilden. Dem.	Hayes. Rep.	Cooper. Gr.	Garfield. Rep.	Hancock. Dem.	Weaver. Gr.
Alabama	102,002	68,230	......	56,178	90,687	4,642
Arkansas	58,071	38,669	289	41,661	60,489	4,079
California	76,465	79,269	47	80,348	80,426	3,392
Colorado	Electors	by Legis	......	27,450	24,647	1,435
Connecticut	61,934	59,034	774	67,073	64,417	868
Delaware	13,381	10,752	......	14,150	15,183	120
Florida	22,923	23,849	......	23,654	27,964	......
Georgia	130,088	50,446	......	52,648	102,522	481
Illinois	258,601	278,232	17,233	318,037	277,321	26,358
Indiana	213,526	208,011	9,533	232,164	225,528	12,986
Iowa	112,099	171,327	9,001	183,904	105,845	32,327
Kansas	37,902	78,322	7,776	121,520	59,789	19,710
Kentucky	159,690	97,156	1,944	104,550	147,999	11,498
Louisiana	70,508	75,135	......	31,891	65,310	422
Maine*	49,823	66,300	663	74,039	65,171	4,408
Maryland	91,780	71,981	33	78,515	93,706	818
Massachusetts	108,777	150,063	779	165,205	111,960	4,548
Michigan	141,095	166,534	9,060	185,190	131,300	34,795
Minnesota	48,799	72,962	2,311	93,903	53,315	3,267
Mississippi	112,173	52,605	......	34,854	75,750	5,797
Missouri	203,077	145,029	3,498	153,567	208,609	35,045
Nebraska	17,554	31,916	2,320	54,979	28,523	3,853
Nevada	9,308	10,383	......	8,732	9,611	......
N. Hampshire	38,509	41,539	76	44,852	40,794	528
New Jersey	115,962	103,517	712	120,555	122,565	2,617
New York	521,949	489,207	1,987	555,544	534,511	12,373
North Carolina	125,427	108,417	......	115,878	124,204	1,136
Ohio	323,182	330,698	3,057	375,048	340,821	6,456
Oregon	14,149	15,206	510	20,619	19,948	249
Pennsylvania	366,158	384,122	7,187	444,704	407,428	20,668
Rhode Island	10,712	15,787	68	18,195	10,779	236
South Carolina	90,906	91,870	......	58,071	112,312	566
Tennessee	133,166	89,566	......	107,677	128,191	5,916
Texas	104,755	44,800	......	57,845	156,228	27,405
Vermont	20,254	44,092	......	45,090	18,181	1,212
Virginia	139,670	95,558	......	84,020	127,976	139
West Virginia	56,455	42,698	1,373	46,243	57,391	9,079
Wisconsin	123,927	130,668	1,509	144,397	114,634	7,980
Total	4,284,757	4,033,950	81,740	4,442,950	4,442,035	307,409

*Weaver's straight vote is 307,409. Counting the full "Fusion" vote of Maine it is 372,743.

OFFICIAL RETURNS.—OHIO.

[Taken from the American Almanac for 1881.]

Ohio By Counties.	President—1880.			Ohio By Counties.	President—1880.		
	Garfield. Rep.	Hancock. Dem.	Weaver. Gr.		Garfield. Rep.	Hancock. Dem.	Weaver. Gr.
Franklin	9,438	9,863	97	Montgomery	9,726	10,332	54
Fulton	2,912	1,787	85	Morgan	2,510	2,091	26
Gallia	3,488	2,310	7	Morrow	2,581	2,143	42
Geauga	3,053	815	52	Muskingum	5,804	5,336	68
Greene	4,927	2,455	9	Noble	2,316	2,044	188
Guernsey	3,318	2,568	26	Ottawa	1,510	2,559	59
Hamilton	35,173	30,122	102	Paulding	1,527	1,431	5
Hancock	3,124	3,350	33	Perry	2,676	3,187	377
Hardin	3,472	3,032	1	Pickaway	2,910	3,753	4
Harrison	2,767	2,082	21	Pike	1,756	2,192	25
Henry	1,738	2,871	26	Portage	3,990	3,147	86
Highland	3,648	3,490		Preble	3,183	2,711	4
Hocking	1,830	2,422	88	Putnam	1,851	3,417	24
Holmes	1,370	3,281	5	Richland	4,032	4,885	6
Huron	4,566	3,040	177	Ross	4,734	4,551	28
Jackson	2,763	2,031	35	Sandusky	3,059	3,640	148
Jefferson	4,434	2,945	77	Scioto	3,639	2,912	61
Knox	3,432	3,475	58	Seneca	4,008	4,845	109
Lake	2,978	1,104	108	Shelby	2,274	3,320	8
Lawrence	4,627	2,862	17	Stark	7,264	6,965	193
Licking	4,210	5,575	74	Summit	5,890	4,071	192
Logan	3,739	2,468	21	Trumbull	6,796	3,184	208
Lorain	5,609	2,752	67	Tuscarawas	4,096	4,844	71
Lucas	7,157	5,985	426	Union	3,302	2,236	2
Madison	2,680	2,305	15	Van Wert	2,634	2,571	5
Mahoning	4,943	4,044	241	Vinton	1,700	1,992	5
Marion	2,192	2,932	7	Warren	4,565	2,564	5
Medina	3,340	2,158	21	Washington	4,711	4,452	112
Meigs	4,103	2,749	10	Wayne	4,424	4,819	23
Mercer	1,473	3,367	13	Williams	2,881	2,596	176
Miami	4,928	3,604	40	Wood	4,305	3,441	179
Monroe	1,600	3,751	106	Wyandot	2,308	2,981	2
Total					375,048	340,821	6,456

OFFICIAL RETURNS.—INDIANA.

[Taken from the American Almanac for 1881.]

INDIANA By Counties.	President—1880.			INDIANA By Counties.	President—1880.		
	Hancock. Dem.	Garfield. Rep.	Weaver. Gr.		Hancock. Dem.	Garfield. Rep.	Weaver. Gr.
Adams	2,226	1,014	21	Johnson	2,461	2,020	287
Allen	7,791	4,815	84	Knox	3,443	2,693	24
Bartholomew	2,730	2,575	57	Kosciusko	2,837	3,571	93
Benton	1,272	1,522	62	Lagrange	1,393	2,567	116
Blackford	1,029	781	127	Lake	1,198	2,102	39
Boone	2,742	2,770	690	Laporte	3,580	3,631	121
Brown	1,576	599	42	Lawrence	1,701	2,057	146
Carroll	2,215	2,205	61	Madison	3,722	2,798	93
Cass	3,579	3,387	119	Marion	11,362	13,803	708
Clark	3,659	2,899	34	Marshall	2,679	2,136	555
Clay	2,893	2,851	363	Martin	1,621	1,311	37
Clinton	3,015	2,565	110	Miami	3,066	2,016	107
Crawford	1,368	1,134	55	Monroe	1,682	1,780	165
Daviess	2,387	2,320	85	Montgomery	3,405	3,643	163
Dearborn	3,615	2,547	27	Morgan	2,046	2,391	133
Decatur	2,291	2,599	94	Newton	716	1,202	103
DeKalb	2,582	2,441	110	Noble	2,878	2,878	81
Delaware	1,826	3,683	59	Ohio	588	727	18
Dubois	2,498	900	15	Orange	1,521	1,421	97
Elkhart	3,472	4,191	187	Owen	1,977	1,486	106
Fayette	1,230	1,760	11	Parke	1,875	1,672	236
Floyd	3,160	2,114	176	Perry	1,867	1,659	27
Fountain	2,261	2,257	554	Pike	1,760	1,618	229
Franklin	3,151	1,683	2	Porter	1,578	2,243	117
Fulton	1,804	1,757	51	Posey	2,615	2,127	23
Gibson	2,477	2,662	74	Pulaski	1,004	897	289
Grant	2,378	3,138	158	Putnam	2,850	2,539	119
Greene	2,246	2,456	192	Randolph	2,058	4,295	44
Hamilton	2,093	3,638	166	Ripley	2,470	2,399	12
Hancock	2,273	1,722	125	Rush	2,324	2,677	52
Harrison	2,481	1,950	131	Scott	1,100	771	16
Hendricks	1,994	3,196	218	Shelby	3,555	2,648	68
Henry	2,031	3,784	253	Spencer	2,475	2,363	79
Howard	1,796	3,000	121	Stark	563	381	178
Huntington	2,657	2,638	125	St. Joseph	3,682	4,147	330
Jackson	3,138	1,997	67	Steuben	1,283	2,325	106
Jasper	848	1,320	91	Sullivan	3,049	1,607	140
Jay	2,161	2,243	156	Switzerland	1,429	1,549	160
Jefferson	2,647	3,296	60	Tippecanoe	3,820	5,061	136
Jennings	1,710	2,068	56	Tipton	1,856	1,518	62

INDIANA—CONTINUED.

INDIANA BY COUNTIES.	PRESIDENT—1880.			INDIANA BY COUNTIES.	PRESIDENT—1880.		
	Hancock. Dem.	Garfield. Rep.	Weaver. Gr.		Hancock. Dem.	Garfield. Rep.	Weaver. Gr.
Union	816	1,085	3	Warrick	2,344	2,008	72
Vanderburgh	4,481	4,909	235	Washington	2,400	1,709	25
Vermillion	1,235	1,562	149	Wayne	3,325	6,252	138
Vigo	4,576	4,983	781	Wells	2,395	1,515	513
Wabash	2,339	3,739	56	White	1,591	1,610	124
Warren	901	1,850	124	Whitney	2,229	1,941	93
Total					225,528	232,164	12,986

OFFICIAL RETURNS—ILLINOIS.

[Taken from the American Almanac for 1881.]

ILLINOIS BY COUNTIES.	PRESIDENT—1880.			ILLINOIS BY COUNTIES.	PRESIDENT—1880.		
	Garfield. Rep.	Hancock. Dem.	Weaver. Gr.		Garfield. Rep.	Hancock. Dem.	Weaver. Gr.
Adams	4,987	6,113	608	Edgar	2,834	2,989	127
Alexander	1,579	1,353	46	Edwards	1,177	575	10
Bond	1,711	1,273	108	Effingham	1,361	2,452	100
Boone	2,038	321	84	Fayette	2,135	2,633	207
Brown	1,008	1,655	153	Ford	1,857	780	455
Bureau	4,099	2,655	329	Franklin	1,286	1,610	283
Calhoun	505	946	22	Fulton	4,108	4,718	553
Carroll	2,396	960	154	Gallatin	1,050	1,574	20
Cass	1,262	1,778	224	Greene	1,865	3,160	49
Champaign	4,720	3,472	566	Grundy	2,087	1,135	202
Christian	2,687	3,346	194	Hamilton	1,002	1,760	499
Clark	1,999	2,374	337	Hancock	3,610	3,957	274
Clay	1,555	1,660	135	Hardin	484	765	10
Clinton	1,578	2,242	116	Henderson	1,270	923	152
Coles	2,991	2,905	141	Henry	4,469	2,961	730
Cook	54,816	44,302	1,168	Iroquois	4,128	2,738	443
Crawford	1,541	1,917	24	Jackson	2,152	2,160	493
Cumberland	1,365	1,563	92	Jasper	1,194	1,761	88
DeKalb	4,124	1,578	104	Jefferson	1,700	2,304	311
DeWitt	2,011	1,845	168	Jersey	1,348	2,107	123
Douglass	1,918	1,689	65	Jo Daviess	2,994	2,363	168
Du Page	2,327	1,229	16	Johnson	1,521	893	170

ILLINOIS—CONTINUED.

Illinois By Counties.	President—1880. Garfield. Rep.	Hancock. Dem.	Weaver. Gr.	Illinois By Counties.	President—1880. Garfield. Rep.	Hancock. Dem.	Weaver. Gr.
Kane	6,180	2,831	410	Piatt	1,855	1,578	156
Kankakee	3,201	1,640	107	Pike	2,968	3,812	777
Kendall	1,954	679	233	Pope	1,561	914	39
Knox	4,863	2,392	869	Pulaski	1,174	742	37
Lake	2,884	1,494	59	Putnam	704	503	2
La Salle	6,941	6,308	892	Randolph	2,705	2,614	41
Lawrence	1,492	1,497	39	Richland	1,628	1,736	2
Lee	3,359	2,242	195	Rock Island	4,025	2,665	1,001
Livingston	3,771	2,861	865	Saline	1,488	1,608	25
Logan	2,729	2,687	121	Sangamon	5,476	6,196	238
Macon	3,447	3,069	185	Schuyler	1,520	1,937	69
Macoupin	3,904	4,341	113	Scott	1,035	1,288	129
Madison	5,024	4,677	115	Shelby	2,017	3,328	1,017
Marion	2,060	2,507	471	Stark	1,382	681	380
Marshall	1,684	1,603	107	St. Clair	5,847	5,877	251
Mason	1,616	1,926	148	Stephenson	3,581	3,071	65
Massac	1,484	778	14	Tazewell	2,919	3,367	153
McDonough	3,014	2,877	468	Union	1,139	2,264	10
McHenry	3,516	1,799	194	Vermillion	4,982	3,421	453
McLean	7,317	5,202	317	Wabash	939	1,142	39
Menard	994	1,478	470	Warren	2,849	2,003	305
Mercer	2,348	1,487	448	Washington	2,280	1,912	44
Monroe	1,172	1,712		Wayne	2,063	2,204	159
Montgomery	2,702	3,173	201	White	1,811	2,591	265
Morgan	3,199	3,452	297	Whiteside	3,918	2,215	403
Moultrie	1,233	1,593	197	Will	5,776	3,803	882
Ogle	4,054	2,085	249	Williamson	1,853	1,825	141
Peoria	5,105	5,705	720	Winnebago	4,617	1,511	278
Perry	1,751	1,535	64	Woodford	2,007	2,364	108
Total					318,037	277,321	26,358

OFFICIAL RETURNS—MISSOURI.

[Taken from the American Almanac for 1881.]

MISSOURI By Counties.	President—1880. Hancock. Dem.	Garfield. Rep.	Weaver. Gr.	MISSOURI By Counties.	President—1880. Hancock. Dem.	Garfield. Rep.	Weaver. Gr.
Adair	1,269	1,657	329	Harrison	1,586	2,097	239
Andrew	1,571	1,781	121	Henry	2,821	1,694	306
Atchison	1,261	1,228	490	Hickory	436	675	252
Audrain	2,322	983	530	Holt	1,297	1,605	212
Barry	1,163	970	327	Howard	2,047	1,166	513
Barton	942	519	712	Howell	726	457	205
Bates	2,949	1,897	245	Iron	854	565	
Benton	962	1,204	164	Jackson	6,703	5,123	732
Bollinger	1,068	629	117	Jasper	2,533	2,874	1,114
Boone	3,269	1,170	418	Jefferson	2,012	1,501	69
Buchanan	4,693	3,317	391	Johnson	2,795	2,400	318
Butler	746	275	96	Knox	1,468	574	765
Caldwell	1,139	1,369	373	Laclede	960	365	774
Callaway	3,369	1,184	110	Lafayette	3,163	1,822	102
Camden	507	563	197	Lawrence	1,476	1,567	337
CapeGirardeau	1,869	1,641	102	Lewis	1,928	1,152	152
Carroll	2,404	2,039	409	Lincoln	2,039	790	634
Carter	238	80	50	Linn	2,049	1,991	182
Cass	2,710	1710	275	Livingston	1,859	1,165	1,268
Cedar	900	926	258	McDonald	706	213	471
Chariton	2,899	1,617	548	Macon	2,880	1,726	844
Christian	438	791	529	Madison	952	391	1
Clarke	1,570	1,503	120	Maries	924	288	58
Clay	2,969	589	193	Marion	3,086	1,811	87
Clinton	2,061	1,237	187	Mercer	990	1,573	231
Cole	1,384	1,338	55	Miller	757	970	167
Cooper	2,189	1,730	372	Mississippi	1,137	525	113
Crawford	1,099	805	69	Moniteau	1,323	853	643
Dade	902	1,237	238	Monroe	3,488	671	120
Dallas	487	634	555	Montgomery	1,721	1,329	343
Daviess	2,045	1,796	285	Morgan	950	798	57
De Kalb	1,305	1,238	221	New Madrid	1,070	341	
Dent	1,073	707	35	Newton	1,535	957	971
Douglas	163	497	556	Nodaway	2,485	2,301	941
Dunklin	1,333	182		Oregon	809	85	23
Franklin	2,260	2,647	78	Osage	1,137	1,117	10
Gasconade	487	1,512		Ozark	314	409	132
Gentry	1,982	1,377	334	Pemiscot	749	85	
Greene	1,912	2,198	1,286	Perry	1,110	887	71
Grundy	1,102	1,917	124	Pettis	2,908	2,457	306

MISSOURI—CONTINUED.

Missouri By Counties.	President—1880.			Missouri By Counties.	President—1880.		
	Hancock. Dem.	Garfield. Rep.	Weaver. Gr.		Hancock. Dem.	Garfield. Rep.	Weaver. Gr.
Phelps	1,132	416	548	Schuyler	1,065	570	457
Pike	3,236	2,151	289	Scotland	1,405	689	479
Platte	2,693	935	49	Scott	1,330	459	....
Polk	1,360	1,506	250	Shannon	467	65	9
Pulaski	772	462	19	Shelby	1,770	350	847
Putnam	725	1,513	424	Stoddard	1,541	590	92
Ralls	1,800	603	14	Stone	140	435	136
Randolph	2,927	1,051	691	Sullivan	1,717	1,693	187
Ray	2,614	908	568	Taney	1,313	337	207
Reynolds	747	39	....	Texas	1,250	477	285
Ripley	578	115	70	Vernon	2,338	940	360
St. Charles	2,191	2,223	33	Warren	662	1,343	203
St. Clair	963	765	1,053	Washington	1,489	775	78
St. Francois	1,750	778	60	Wayne	1,144	568	46
Ste. Genevieve	1,081	650	40	Webster	1,024	561	616
St. Louis Co.	2,719	3,223	4	Worth	751	657	163
St. Louis City	23,837	23,006	872	Wright	409	641	365
Saline	3,851	1,907	359				
Total					208,609	153,567	35,045

POPULATION OF THE UNITED STATES BY RACES IN 1880.

From the official returns of the tenth census, subject to final correction.

[From the American Almanac, 1881.]

STATES AND TERRITORIES.	Total Population 1880.	White. 1880.	Colored. 1880.	Chinese 1880.	Indians. civ'd or taxed. 1880.
Alabama	1,262,794	661,986	600,141	4	213
Arizona	40,441	35,178	158	1,632	3,493
Arkansas	802,564	591,611	210,622	134	197
California	864,686	767,266	6,168	75,122	16,130
Colorado	194,649	191,452	2,459	610	128
Connecticut	622,683	610,884	11,422	130	241
Dakota	135,180	133,177	381	238	1,384
Delaware	146,654	120,198	26,456		
District of Columbia	177,638	118,236	59,378	18	6
Florida	267,351	141,249	125,262	18	37
Georgia	1,539,048	814,218	724,654	17	94
Idaho	32,611	29,011	58	3,378	164
Illinois	3,078,769	3,032,174	46,248	214	133
Indiana	1,978,362	1,939,094	38,998	37	233
Iowa	1,624,620	1,614,510	9,442	47	464
Kansas	995,966	952,056	43,096	22	792
Kentucky	1,648,708	1,377,077	271,462	10	50
Louisiana	940,103	455,063	483,898	483	819
Maine	648,945	646,903	1,427	8	607
Maryland	934,632	724,718	209,897	6	11
Massachusetts	1,783,012	1,764,082	18,644	256	341
Michigan	1,636,331	1,614,078	14,986	29	7,238
Minnesota	780,806	776,940	1,558	54	2,254
Mississippi	1,131,592	479,371	650,337	52	1,832
Missouri	2,168,804	2,023,568	145,046	94	96
Montana	39,157	35,468	202	1,737	1,750
Nebraska	452,433	449,805	2,376	18	233
Nevada	62,265	53,574	465	5,423	2,803
New Hampshire	346,984	346,264	646	14	60
New Jersey	1,130,983	1,091,856	38,796	182	59
New Mexico	118,430	107,188	907	55	10,280
New York	5,083,810	5,017,142	64,943	942	783
North Carolina	1,400,047	867,467	531,316	1	1,216
Ohio	3,198,239	3,118,344	79,665	117	113
Oregon	174,767	163,087	493	9,508	1,679
Pennsylvania	4,282,784	4,197,106	85,342	170	168
Rhode Island	276,528	269,933	6,503	27	67
South Carolina	995,622	391,258	604,325	9	114
Tennessee	1,542,463	1,139,120	402,992	26	326
Texas	1,592,574	1,197,493	394,007	142	932
Utah	143,906	142,381	204	518	804
Vermont	332,286	331,243	1,032		11
Virginia	1,512,806	880,739	631,996	6	65
Washington	75,120	67,349	357	3,227	4,187
West Virginia	618,443	592,433	25,729	14	17
Wisconsin	1,315,480	1,309,622	2,724	16	3,118
Wyoming	20,788	19,436	299	914	139
Total. United States	50,152,866	43,402,408	6,577,497	105,679	65,880

OFFICIAL RETURNS OF THE CENSUS FOR 1880.—OHIO.

[From the American Almanac for 1881.]

OHIO BY COUNTIES.	Population. 1870.	Population. 1880.	OHIO BY COUNTIES.	Population. 1870.	Population. 1880.
Adams	20,750	24,004	Licking	35,756	40,451
Allen	23,623	31,323	Logan	23,028	26,268
Ashland	21,933	23,883	Lorain	30,308	35,525
Ashtabula	32,517	37,139	Lucas	46,722	67,388
Athens	23,768	28,451	Madison	15,633	20,129
Auglaise	20,041	25,443	Mahoning	31,001	42,867
Belmont	39,714	49,638	Marion	16,184	20,564
Brown	30,802	32,726	Medina	20,092	21,454
Butler	39,912	42,580	Meigs	31,465	32,325
Carroll	14,491	16,416	Mercer	17,254	21,808
Campaign	24,188	27,817	Miami	32,740	36,178
Clark	32,070	41,947	Monroe	25,779	26,497
Clermont	34,268	36,713	Montgomery	64,006	78,545
Clinton	21,914	27,539	Morgan	20,363	20,074
Columbiana	38,299	48,603	Morrow	18,583	19,073
Coshocton	23,600	26,640	Muskingum	44,886	49,612
Crawford	25,556	30,583	Noble	19,949	21,137
Cuyahoga	132,010	196,937	Ottawa	13,364	19,763
Darke	32,278	40,498	Paulding	8,544	13,489
Defiance	15,719	22,518	Perry	18,453	28,218
Delaware	25,175	27,380	Pickaway	24,875	27,353
Erie	28,188	32,640	Pike	15,447	17,927
Fairfield	31,138	34,283	Portage	24,584	27,500
Fayette	17,170	20,364	Preble	21,809	24,534
Franklin	63,019	86,816	Putnam	17,081	23,718
Fulton	17,789	21,062	Richland	32,516	36,305
Gallia	25,545	28,124	Ross	37,097	40,307
Geauga	14,190	14,255	Sandusky	25,503	32,063
Greene	28,038	31,348	Scioto	29,302	33,504
Guernsey	23,838	27,197	Seneca	30,827	36,955
Hamilton	260,370	313,345	Shelby	20,748	24,136
Hancock	23,847	27,788	Stark	52,508	64,027
Hardin	18,714	27,028	Summit	34,674	43,788
Harrison	18,682	20,455	Trumbull	38,659	44,852
Henry	14,028	20,587	Tuscarawas	33,840	40,197
Highland	29,133	30,277	Union	18,730	22,374
Hocking	17,925	21,126	Van Wert	15,823	23,030
Holmes	18,177	20,775	Vinton	15,027	17,226
Huron	28,532	31,609	Warren	26,689	28,392
Jackson	21,759	23,679	Washington	40,609	43,264
Jefferson	29,188	33,018	Wayne	35,116	37,452
Knox	26,333	27,450	Williams	20,991	23,821
Lake	15,935	16,326	Wood	24,596	34,026
Lawrence	31,380	39,068	Wyandot	18,553	22,401
Total				2,665,260	3,198,239

INDIANA.

OFFICIAL RETURNS OF THE CENSUS FOR 1880.

[From the American Almanac for 1881.]

Indiana by Counties.	Population 1870.	Population 1880.	Indiana by Counties.	Population 1870.	Population 1880.
Adams	11,382	15,385	Lawrence	14,628	18,453
Allen	43,494	54,765	Madison	22,770	27,531
Bartholomew	21,133	22,777	Marion	71,939	102,780
Benton	5,615	11,107	Marshall	20,211	23,416
Blackford	6,272	8,021	Martin	11,103	13,474
Boone	22,593	25,921	Miami	21,052	24,081
Brown	8,681	10,264	Monroe	14,168	15,874
Carroll	16,152	18,347	Montgomery	23,765	27,314
Cass	24,193	27,609	Morgan	17,528	18,899
Clarke	24,770	28,638	Newton	5,829	8,167
Clay	19,084	25,853	Noble	20,389	23,007
Clinton	17,330	23,473	Ohio	5,837	5,563
Crawford	9,851	12,356	Orange	13,497	14,363
Daviess	16,747	21,552	Owen	16,137	15,901
Dearborn	24,116	26,656	Parke	18,166	19,460
Decatur	19,053	19,779	Perry	14,801	16,997
De Kalb	17,167	20,223	Pike	13,779	16,384
Delaware	19,030	22,927	Porter	13,942	17,229
Dubois	12,597	15,991	Posey	19,185	20,857
Elkhart	26,026	33,453	Pulaski	7,801	9,851
Fayette	10,476	11,394	Putnam	21,514	22,502
Floyd	23,300	24,589	Randolph	22,862	26,437
Fountain	16,389	20,228	Ripley	20,977	21,627
Franklin	20,223	20,090	Rush	17,626	19,238
Fulton	12,726	14,301	St. Joseph	25,322	33,176
Gibson	17,371	22,742	Scott	7,873	8,343
Grant	18,487	23,618	Shelby	21,892	25,256
Greene	19,514	22,996	Spencer	17,998	22,122
Hamilton	20,882	24,809	Starke	3,888	5,105
Hancock	15,123	11,123	Steuben	12,854	14,644
Harrison	19,913	21,326	Sullivan	18,453	20,433
Hendricks	20,277	22,975	Switzerland	12,134	13,336
Henry	22,986	24,015	Tippecanoe	33,515	35,966
Howard	15,847	19,584	Tipton	11,953	14,402
Huntington	19,036	21,805	Union	6,341	7,673
Jackson	18,974	23,050	Vanderburgh	33,145	42,192
Jasper	6,354	9,465	Vermilion	10,840	12,025
Jay	15,000	19,282	Vigo	33,549	45,656
Jefferson	29,741	25,977	Wabash	21,305	25,240
Jennings	16,218	16,453	Warren	10,204	11,497
Johnson	18,366	19,537	Warwick	17,653	20,162
Knox	21,562	26,323	Washington	18,495	18,949
Kosciusko	23,531	26,492	Wayne	34,048	38,614
La Grange	14,148	15,629	Wells	13,585	18,442
Lake	12,339	15,091	White	10,554	13,793
La Porte	27,062	30,976	Whitley	14,399	16,941
Total				1,680,637	1,978,362

ILLINOIS.

OFFICIAL RETURN OF THE TENTH CENSUS.

[From the American Almanac for 1881.]

Illinois by Counties.	Population 1870.	Population 1880.	Illinois by Counties.	Population 1870.	Population 1880.
Adams	56,362	59,148	Lee	27,171	27,491
Alexander	10,564	14,809	Livingston	31,471	38,450
Bond	13,152	14,873	Logan	23,053	25,041
Boone	12,942	11,527	McDonough	26,509	27,984
Brown	12,205	13,044	McHenry	23,762	24,914
Bureau	32,415	33,189	McLean	53,988	60,115
Calhoun	6,562	7,471	Macon	26,481	30,671
Carroll	16,705	16,895	Macoupin	32,726	37,705
Cass	11,580	14,494	Madison	44,131	50,141
Champaign	32,737	40,869	Marion	20,622	23,691
Christian	20,363	28,232	Marshall	16,956	15,036
Clark	18,719	21,900	Mason	16,184	16,244
Clay	15,875	16,195	Massac	9,581	10,443
Clinton	16,285	18,718	Menard	11,735	13,028
Coles	25,235	27,055	Mercer	18,769	19,501
Cook	349,966	607,468	Monroe	12,982	13,682
Crawford	13,889	16,190	Montgomery	25,314	28,086
Cumberland	12,223	13,762	Morgan	28,463	31,519
DeKalb	23,265	26,774	Moultrie	10,385	13,705
DeWitt	14,768	17,014	Ogle	27,492	29,946
Douglass	13,484	15,857	Peoria	47,540	55,419
Du Page	16,685	19,187	Perry	13,723	16,008
Edgar	21,450	25,504	Piatt	10,953	15,583
Edwards	7,565	8,600	Pike	30,768	33,761
Effingham	15,653	18,924	Pope	11,437	13,256
Fayette	19,638	23,243	Pulaski	8,752	9,507
Ford	9,103	15,105	Putnam	6,280	5,555
Franklin	12,652	16,129	Randolph	20,859	25,691
Fulton	38,291	41,249	Richland	12,803	15,546
Gallatin	11,134	12,862	Rock Island	29,783	38,314
Greene	20,277	23,014	St Clair	51,068	61,850
Grundy	14,938	16,738	Saline	12,714	15,940
Hamilton	13,014	16,712	Sangamon	46,352	52,902
Hancock	35,935	35,352	Schuyler	17,419	16,249
Hardin	5,113	6,024	Scott	10,530	10,745
Henderson	12,582	10,755	Shelby	25,476	30,282
Henry	35,506	36,609	Stark	10,751	11,209
Iroquois	25,782	35,457	Stephenson	30,608	31,970
Jackson	19,634	22,508	Tazewell	27,903	29,679
Jasper	11,234	14,515	Union	16,518	18,100
Jefferson	17,864	20,686	Vermillion	30,388	41,600
Jersey	15,054	15,546	Wabash	8,841	9,945
Jo Daviess	27,820	27,534	Warren	23,174	22,940
Johnson	11,248	13,079	Washington	17,599	21,117
Kane	39,091	44,956	Wayne	19,758	21,297
Kankakee	24,352	24,961	White	16,846	23,089
Kendall	12,399	13,084	Whitesides	27,503	30,888
Knox	39,522	38,360	Will	43,013	53,424
Lake	21,014	21,299	Williamson	17,329	19,326
LaSalle	60,792	70,420	Winnebago	29,301	30,518
Lawrence	12,533	13,663	Woodford	18,956	21,630
Total				2,539,891	3,078,769

OFFICIAL RETURN OF THE TENTH CENSUS—MISSOURI.

[From the American Almanac for 1881.]

Missouri By Counties.	Population 1870.	Population 1880.	Missouri By Counties.	Population 1870.	Population 1880.
Adair	11,448	15,190	Livingston	16,730	20,205
Andrew	15,137	16,318	Macon	23,230	16,223
Atchison	8,440	14,565	Madison	5,849	8,860
Audrain	12,307	19,780	Maries	5,916	7,323
Barry	10,373	14,434	Marion	23,780	24,837
Barton	5,087	10,332	McDonald	5,226	7,816
Bates	15,960	25,382	Mercer	11,557	14,674
Benton	11,322	12,398	Miller	6,616	9,807
Bollinger	8,162	11,132	Mississippi	4,982	9,270
Boone	20,765	25,444	Moniteau	11,375	14,349
Buchanan	35,109	49,820	Monroe	17,149	19,075
Butler	4,298	6,011	Montgomery	10,405	16,251
Caldwell	11,390	13,654	Morgan	8,434	10,134
Callaway	19,202	23,670	New Madrid	6,357	7,694
Camden	6,108	7,267	Newton	12,821	18,948
Cape Girardeau	17,558	20,998	Nodaway	14,751	29,560
Carroll	17,446	23,262	Oregon	3,287	5,791
Carter	1,455	2,168	Osage	10,793	11,824
Cass	19,296	22,431	Ozark	3,363	5,618
Cedar	9,474	10,757	Pemiscot	2,059	4,299
Chariton	19,136	25,224	Perry	9,877	11,895
Christian	6,707	9,649	Pettis	18,706	27,298
Clarke	13,667	15,031	Phelps	10,506	12,565
Clay	15,564	15,579	Pike	23,076	26,716
Clinton	14,063	16,073	Platte	17,352	17,373
Cole	10,292	15,519	Polk	12,445	15,745
Cooper	20,692	21,638	Pulaski	4,714	7,250
Crawford	7,982	10,774	Putnam	11,217	13,556
Dade	8,683	12,557	Ralls	10,500	11,838
Dallas	8,383	9,272	Randolph	15,908	22,751
Daviess	14,410	19,174	Ray	18,700	20,200
De Kalb	9,858	13,344	Reynolds	3,756	5,722
Dent	6,357	10,647	Ripley	3,175	5,377
Dodge			Rives		
Douglass	3,915	7,753	Saline	21,672	29,938
Dunklin	5,982	9,604	Schuyler	8,820	10,470
Franklin	30,098	26,536	Scotland	10,670	12,507
Gasconade	10,093	11,173	Scott	7,317	8,587
Gentry	11,607	17,202	Shannon	2,339	3,441
Greene	21,549	28,839	Shelby	10,119	14,024
Grundy	10,567	15,210	St. Charles	21,304	23,060
Harrison	14,685	20,318	St. Clair	6,742	14,157
Henry	17,401	23,343	St. Francois	9,742	13,821
Hickory	6,452	7,388	Ste. Genevieve	8,384	10,390
Holt	11,652	15,510	St. Louis (city)	351,189	350,522
Howard	17,233	18,428	St. Louis		31,888
Howell	4,218	8,814	Stoddard	8,535	13,432
Iron	6,278	8,183	Stone	3,253	4,429
Jackson	55,041	82,364	Sullivan	11,907	16,569
Jasper	14,928	32,021	Taney	4,407	5,633
Jefferson	15,380	18,736	Texas	9,618	12,219
Johnson	24,648	28,177	Van Buren		
Knox	10,974	13,047	Vernon	11,247	10,382
Laclede	9,380	11,524	Warren	9,673	10,806
Lafayette	22,623	25,750	Washington	11,719	12,895
Lawrence	13,067	17,585	Wayne	6,068	9,097
Lewis	15,114	15,925	Webster	10,434	12,176
Lincoln	15,960	17,443	Worth	5,004	8,208
Linn	15,900	20,016	Wright	5,684	9,733
Total				1,721,295	2,168,804

KANSAS.

OFFICIAL RETURNS ON PRESIDENTIAL VOTE FOR 1880.

[From the American Almanac, 1881.]

KANSAS BY COUNTIES.	Garfield. Rep.	Hancock. Dem.	Weaver. Gr.	KANSAS By Counties.	Garfield. Rep.	Hancock. Dem.	Weaver. Gr.
Allen	1,576	803	44	Linn	1,990	745	577
Anderson	1,127	497	370	Lyon	2,398	869	402
Atchison	2,834	2,132	71	Marion	1,239	539	271
Barbour	262	175	63	Marshall	2,276	997	427
Barton	1,172	714	62	McPherson	2,225	564	545
Bourbon	2,320	1,161	364	Miami	2,010	1,324	454
Brown	1,850	896	107	Mitchell	1,728	797	235
Butler	2,398	1,119	433	Montgomery	1,773	1,294	693
Chase	716	324	409	Morris	1,281	550	179
Chautauqua	1,321	655	333	Nemaha	1,755	934	5
Cherokee	2,374	1,681	855	Neosho	1,471	948	461
Clay	1,765	531	369	Ness	315	129	....
Cloud	2,156	888	65	Norton	761	337	198
Coffey	1,422	851	189	Osage	2,704	907	793
Cowley	2,630	1,557	190	Osborne	1,446	589	61
Crawford	1,902	1,356	450	Ottawa	1,443	524	333
Davis	702	399	335	Pawnee	697	235	17
Decatur	307	163	28	Phillips	1,261	553	221
Dickinson	1,954	886	292	Pottawatomie	2,138	1,179	224
Doniphan	2,067	1,143	51	Pratt	196	97	34
Douglas	3,048	1,463	246	Reno	1,384	536	252
Edwards	312	102	....	Republic	1,875	661	151
Elk	1,274	458	486	Rice	1,108	496	313
Ellis	680	420	54	Riley	1,484	376	347
Ellsworth	1,077	483	32	Rooks	805	338	326
Ford	370	288	....	Rush	542	238	24
Franklin	2,108	728	898	Russell	922	317	110
Graham	494	104	210	Saline	1,950	838	95
Greenwood	1,311	667	347	Sedgwick	2,288	1,354	364
Harper	546	294	170	Shawnee	4,403	1,548	123
Harvey	1,554	585	136	Sheridan	93	52	64
Hodgeman	176	52	13	Smith	1,524	517	406
Jackson	1,504	853	14	Stafford	530	192	60
Jefferson	1,976	1,397	78	Sumner	2,073	4,419	529
Jewell	2,199	883	399	Trego	332	107	29
Johnson	2,132	1,182	354	Wabaunsee	1,279	510	39
Kingman	436	200	85	Washington	1,957	827	230
Labette	2,721	1,462	420	Wilson	1,627	722	527
Leavenworth	3,188	2,489	171	Woodson	898	437	9
Lincoln	957	419	154	Wyandotte	2,410	1,733	236
Total					121,520	59,789	19,710

POPULATION OF KANSAS FOR 1870 AND 1880.

Kansas By Counties	Population 1870	Population 1880	Kansas By Counties	Population 1870	Population 1880
Allen	7,022	11,307	Lincoln	516	8,582
Anderson	5,220	9,059	Linn	12,174	15,299
Arapahoe		3	Lykins		
Atchison	15,507	26,674	Lyon	8,014	17,327
Barbour		2,661	Madison		
Barton	2	10,319	Marion	768	12,457
Bourbon	15,076	19,595	Marshall	6,901	16,135
Breckenridge			McGhee		
Brown	6,823	12,319	McPherson	738	17,143
Buffalo		191	Meade		296
Butler	3,035	18,587	Miami	11,725	17,818
Chase	1,975	6,081	Mitchell	485	14,913
Chautauqua		11,072	Montgomery	7,564	18,230
Cherokee	11,038	21,918	Morris	2,225	9,266
Cheyenne		37	Nemaha	7,339	12,463
Clarke		163	Neosho	10,206	15,124
Clay	2,942	12,320	Ness	2	3,722
Cloud	2,323	15,346	Norton		7,002
Coffey	6,201	11,438	Osage	7,648	19,643
Comanche		372	Osborne	33	12,518
Cowley	1,175	21,539	Otoe		
Crawford	8,160	16,854	Ottawa	2,127	10,308
Davis	5,526	6,994	Pawnee	179	5,396
Decatur		4,180	Phillips		12,017
Dickinson	3,043	14,973	Pottawatamie	7,848	16,347
Doniphan	13,969	14,258	Pratt		1,890
Dorn			Rawlins		1,623
Douglas	20,592	21,706	Reno		12,824
Edwards		2,409	Republic	1,281	14,913
Elk		10,625	Rice	5	9,292
Ellis	1,336	6,179	Riley	5,105	10,430
Ellsworth	1,185	8,494	Rooks		8,113
Foote		411	Rush		5,490
Ford	427	3,122	Russell	156	7,351
Franklin	10,385	16,800	Saline	4,246	13,810
Godfrey			Scott		43
Gove		1,196	Sedgwick	1,095	18,753
Graham		4,258	Sequoyah		568
Grant		9	Seward		5
Greeley		3	Shawnee	13,121	29,092
Greenwood	3,484	10,550	Sheridan		1,567
Hamilton		168	Sherman		13
Harper		4,133	Smith	66	13,885
Harvey		11,454	Stafford		4,755
Hodgman		1,704	Stanton		5
Howard	2,794		Stevens		12
Hunter			Sumner	22	20,812
Jackson	6,053	10,718	Thomas		161
Jefferson	12,526	15,564	Trego	166	2,535
Jewell	207	17,477	Wabaunsee	3,362	8,757
Johnson	13,684	16,886	Wallace	538	686
Kansas		9	Washington	4,081	14,910
Kearney		159	Wichita		14
Kingman		3,713	Wilson	6,694	13,776
Labette	9,973	22,746	Woodson	3,827	6,535
Lane		632	Wyandotte	10,015	19,151
Leavenworth	32,444	31,673			
Total				364,399	995,966

CONGRESSIONAL DISTRICTS OF INDIANA AND THE TOTAL VOTE OF EACH BY COUNTIES, 1880.

FIRST—
- Gibson 5,186
- Perry 3,541
- Pike 3,555
- Posey 4,639
- Spencer 4,882
- Vanderburg 9,667
- Warrick 4,403

    Totals 35,873

SECOND—
- Daviess 4,815
- Dubois 3,375
- Greene 5,054
- Knox 6,090
- Lawrence 3,927
- Martin 2,878
- Orange 3,065
- Sullivan 4,857

    Totals 34,061

THIRD—
- Clarke 6,568
- Crawford 2,611
- Floyd 5,375
- Harrison 4,428
- Jackson 5,196
- Jennings 3,890
- Scott 1,884
- Washington 4,112

    Totals 34,064

FOURTH—
- Dearborn 6,177
- Decatur 4,994
- Franklin 4,825
- Jefferson 6,083
- Ohio 1,355
- Ripley 4,887
- Switzerland 3,199
- Union 1,946

    Totals 33,466

FIFTH—
- Bartholomew 5,562
- Brown 2,169
- Hendricks 5,425
- Johnson 4,752
- Monroe 3,572
- Morgan 4,620
- Owen 3,593
- Putnam 5,503

    Totals 35,186

SIXTH—
- Delaware 5,490
- Fayette 2,986
- Henry 6,031
- Randolph 6,296
- Rush 5,036
- Wayne 9,746

    Totals 35,585

SEVENTH—
- Hancock 4,171
- Marion 26,217
- Shelby 6,163

    Totals 36,551

EIGHTH—
- Clay 6,179
- Fountain 5,164
- Montgomery 7,180
- Parke 4,753
- Vermilion 2,930
- Vigo 10,352
- Warren 2,848

    Totals 39,386

NINTH—
- Boone 6,348
- Clinton 5,630
- Hamilton 5,849
- Madison 6,542
- Tippecanoe 9,059
- Tipton 3,442

    Totals 36,870

TENTH—
- Benton 2,817
- Carroll 4,513
- Cass 7,044
- Fulton 3,605
- Jasper 2,270

CONGRESSIONAL DISTRICTS OF INDIANA, ETC.—CONTINUED.

Lake	3,299	TWELFTH—	
Newton	1,962	Allen	12,454
Porter	3,984	Noble	5,761
Pulaski	2,176	Steuben	3,607
White	3,360	Whitley	4,191
		La Grange	3,735
Totals	35,030	De Kalb	5,082
ELEVENTH—			
Adams	3,164	Totals	34,830
Blackford	1,868	THIRTEENTH—	
Grant	5,701	Elkhart	7,779
Howard	4,890	Kosciusko	6,457
Huntington	5,389	La Porte	7,627
Jay	4,545	Marshall	5,251
Miami	6,133	St. Joseph	8,224
Wabash	6,067	Starke	1,205
Wells	4,370		
		Totals	36,543
Totals	42,127		

CONGRESSIONAL DISTRICTS OF ILLINOIS AND THE TOTAL VOTE OF EACH BY COUNTIES, 1880.

FIRST—		SIXTH—	
Chi. Wards and Towns	37,894	Henry	7,265
Du Page	3,577	Le Bureau	7,082
		Lee	5,781
SECOND— Totals	41,471	Putnam	1,212
Chicago Wards	37,620	Rock Island	7,588
THIRD—			
Chi. Wards and Towns	24,529	SEVENTH— Totals	28,918
Lake	4,477	Grundy	3,424
		Kendall	2,860
FOURTH— Totals	29,006	La Salle	14,133
Boone	2,470	Will	10,486
De Kalb	5,800		
Kane	9,418	EIGHTH— Totals	30,895
McHenry	5,513	Ford	7,442
Winnebago	6,399	Iroquois	7,318
		Kankakee	4,946
FIFTH— Totals	29,295	Livingstone	7,442
Carroll	3,543	Marshall	3,393
Jo Daviess	7,873	Woodford	4,475
Ogle	6,500		
Stephenson	6,724	Totals	30,676
Whitesides	6,549	NINTH—	
		Fulton	9,436
Totals	28,689	Knox	8,180

CONGRESSIONAL DISTRICTS OF ILLINOIS, ETC.—CONTINUED.

Peoria	11,468
Stark	2,430
Tenth— Totals	31,500
Hancock	7,840
Henderson	2,341
Mercer	4,244
McDonough	6,317
Schuyler	3,524
Warren	5,126
Eleventh— Totals	29,388
Adams	11,654
Brown	2,804
Calhoun	1,472
Green	5,053
Jersey	3,571
Pike	7,494
Twelfth— Totals	32,134
Cass	3,260
Christian	6,225
Menard	2,928
Morgan	6,938
Sangamon	11,892
Scott	2,451
Thirteenth— Totals	33,694
De Witt	4,023
Logan	5,590
Mason	3,689
McLean	12,834
Tazewell	6,412
Fourteenth— Totals	32,544
Champaign	8,867
Coles	6,028
Douglas	3,641
Macon	6,724
Piatt	3,572
Vermilion	8,819
Fifteenth— Totals	37,441
Clark	4,718
Crawford	3,464
Cumberland	3,004
Edgar	5,923
Effingham	3,906
Jasper	3,935
Lawrence	3,022
Moultrie	3,002
Shelby	6,098
Sixteenth— Totals	36,173
Bond	3,035
Clay	3,344
Clinton	3,927
Fayette	4,969
Marion	5,026
Montgomery	6,072
Washington	4,228
Seventeenth— Totals	30,643
Macoupin	8,296
Madison	9,841
Monroe	2,877
St. Clair	11,924
Eighteenth— Totals	32,937
Alexander	2,971
Jackson	4,800
Johnson	2,581
Massac	2,174
Perry	3,349
Pope	2,520
Pulaski	1,950
Randolph	5,305
Williamson	3,818
Union	3,413
Nineteenth— Totals	33,019
Edwards	1,767
Franklin	3,151
Gallatin	2,632
Hamilton	3,219
Hardin	1,252
Jefferson	4,319
Richland	3,372
Saline	3,126
Wabash	2,119
Wayne	4,436
White	4,750
Totals	34,038

The following table shows when the next Legislature of each state meets, the length of session, and whether biennial or annual; also the number of counties and the legal rate of interest in each state of the Union:

STATES.	Session.	Limit of Session.	Next Legis-lature Meets	Legal Rate of Interest.	No Counties in State.
Arkansas	Biennial	60 days	Jan. 10, 1883	6%	55
Alabama	Biennial	50 days	Nov. 7, 1882	8%	52
California	Biennial	60 days	Jan. 5, 1882	10%	44
Colorado	Biennial	40 days	Jan. 5, 1883	10%	31
Connecticut	Annual	None	Jan. 5, 1882	6%	8
Delaware	Biennial	None	Jan. 4, 1883	6%	3
Florida	Biennial	60 days	Jan. 4, 1883	8%	37
Georgia	Biennial	40 days	Nov. 3, 1882	7%	132
Illinois	Biennial	None	Jan. 5, 1883	6%	102
Indiana	Biennial	60 days	Jan. 6, 1883	8%	92
Iowa	Biennial	None	Jan. 12, 1882	6%	99
Kansas	Biennial	50 days	Jan. 11, 1883	7%	41
Kentucky	Biennial	60 days	Dec. 31, 1883	6%	109
*Louisiana	Biennial	90 days	Jan. 12, 1882	5%	48
Maine	Biennial	None	Jan. 5, 1883	6%	16
Maryland	Biennial	90 days	Jan. 4, 1882	6%	22
Massachusetts	Annual	None	Jan. 5, 1882	6%	14
Michigan	Biennial	None	Jan. 5, 1883	7%	62
Minnesota	Biennial	60 days	Jan. 4, 1883	7%	64
Mississippi	Biennial	None	Jan. 6, 1882	6%	60
Missouri	Biennial	70 days	Jan. 5, 1883	6%	113
Nebraska	Biennial	40 days	Jan. 4, 1883	10%	62
Nevada	Biennial	60 days	Jan. 3, 1883	10%	14
New Hampshire	Biennial	None	Jan. 1, 1883	6%	10
New Jersey	Annual	None	Jan. 11, 1882	6%	21
New York	Annual	None	Jan. 6, 1882	6%	60
North Carolina	Biennial	60 days	Jan. 5, 1883	6%	87
Ohio	Biennial	None	Jan. 5, 1882	6%	88
Oregon	Biennial	40 days	Jan. 13, 1882	10%	19
Pennsylvania	Biennial	None	Jan. 4, 1883	6%	65
Rhode Island	Annual	None	Jan. 18, 1882	6%	5
†South Carolina	Annual	None	Nov. 23, 1882	7%	30
Tennessee	Biennial	75 days	Jan. 3, 1883	6%	84
Texas	Biennial	60 days	Jan. 3, 1883	8%	151
Vermont	Biennial	None	Oct. 3, 1882	6%	14
Virginia	Biennial	90 days	Dec. 18, 1883	6%	94
West Virginia	Biennial	45 days	Jan. 12, 1883	6%	54
Wisconsin	Annual	None	Jan. 12, 1882	7%	58

*Called Parishes. †Called Districts.

CHIEF JUSTICES OF THE SUPREME COURT OF THE UNITED STATES.

From 1789 to 1881.

John Jay........N. Y.,	served from	1789 to 1795,	6 yrs..	Resigned.
Oliver Ellsworth. Ct.,	" "	1796 to 1801,	5 yrs..	Resigned.
John Marshall... Va.,	" "	1801 to 1835,	34 yrs..	Died.
Roger B. Taney.. Md.,	" "	1836 to 1864,	28 yrs..	Died.
Salmon P. Chase. O.,	" "	1864 to 1873,	9 yrs..	Died.
Morrison R.Waite O.,	" "	1874—		

Present Supreme Court of the United States, and its Associate Justices:

MORRISON R. WAITE, Chief Justice.

ASSOCIATE JUSTICES.

Nathan Clifford......Me.	Commissioned	1857, by Buchanan.
Noah H. Swayne.....O.	"	1862, by Lincoln.
Samuel F. Miller.....Ia.........	"	1862, by Lincoln.
Stephen J. Field .....Cal........	"	1863, by Lincoln.
W. B. Woods ....Ga.	"	1880, by Hayes.
Joseph P. Bradley....N. J.	"	1870, by Grant.
Ward Hunt...........N. Y.	"	1872, by Grant.
John M. Harlan......Ky.	"	1877, by Hayes.
Stanley Mathews....O..........	"	1881, by Garfield.

SENATORS-ELECT TO FORTY-SEVENTH CONGRESS.

[From the American Almanac for 1881.]

Forty-seventh Congress—March 4, 1881, to March 4, 1883.

Alabama.
John T. Morgan, D., Selma............................1883.
James L. Pugh, D., Eufaula............................1885.

Arkansas.
Aug. H. Garland, D., Little Rock......................1883.
James D. Walker, D., Fayetteville......................1885.

California.
James T. Farley, D., Jackson..........................1885.
John F. Miller, R., San Francisco......................1887.

Colorado.
Henry M. Teller, R., Central City......................1883.
Nathaniel P. Hill, R., Denver.. ....1885.

Connecticut.
Orville H. Platt, R., Meriden.....................................1885.
Joseph R. Hawley, R., Hartford ....1887.

Delaware.
Eli Saulsbury, D., Kenton...1883.
Thomas F. Bayard, D., Wilmington ....1887.

Florida.
Wilkinson Call, D., Jacksonville ....1885.
Charles W. Jones, D., Pensacola ....1887.

Georgia.
Benjamin H. Hill, D., Atlanta ....1883.
Joseph E. Brown, D., Atlanta ....1885.

Illinois.
David Davis, Ind. D., Bloomington..............................1883.
John A. Logan, R., Chicago.......................................1885.

Indiana.
Daniel W. Voorhees, D., Terre Haute..........................1885.
Benjamin Harrison, R., Indianapolis ....1887.

Iowa.
J. W. McDill, R...1883.
William B. Allison, R., Dubuque.................................1885.

Kansas.
Preston B. Plumb, R., Emporia ....1883.
John J. Ingalls, R., Atchison.....................................1885.

Kentucky.
James B. Beck, D., Lexington....................................1883.
John S. Williams, D., Mt. Sterling..............................1885.

Louisiana.
William P. Kellogg, R., New Orleans..........................1883.
B. Frank Jonas, D., New Orleans...............................1885.

Maine.
W. P. Frye, R..1883.
Eugene Hale, R., Ellsworth......................................1887.

Maryland.
James B. Groome, D., Elkton....................................1885.
Arthur P. Gorman, D., Laurel...................................1887.

Massachusetts.
George F. Hoar, R., Worcester..................................1885.
Henry L. Dawes, R., Pittsfield ....1887.

Michigan.
Thomas W. Ferry, R., Grand Haven ....1883.
Omar D. Conger, R., Port Huron..........................1887.

Minnesota.
A. J. Edgerton, R., Kasson ....1883.
Sam J. R. McMillan, R., St. Paul ....1887.

Mississippi.
Lucius Q. C. Lamar, D., Oxford ....1883.
James Z. George, D., Carrollton ....1887.

Missouri.
George G. Vest, D., Sedalia..............................1885.
Francis M. Cockrell, D., Warrensburg....................1887.

Nebraska.
Alvin Saunders, R., Omaha................................1883.
Charles H. Van Wyck, R., Nebraska City..................1887.

Nevada.
John P Jones, R., Gold Hill ....1885.
James G. Fair, D., Virginia City.........................1887.

New Hampshire.
Edward H. Rollins, R., Concord ....1883.
Henry W. Blair, R., Plymouth.............................1885.

New Jersey.
John R. McPherson, D., Jersey City ....1883.
William J. Sewell, R., Camden............................1887.

New York.
*Roscoe Conkling, R., Utica ....1885.
*Thomas C. Platt, R., Owego..............................1887.

North Carolina.
Matt. W. Ransom, D., Weldon ....1883.
Zebulon B. Vance, D., Charlotte ....1885.

Ohio.
George H. Pendleton, D., Cincinnati ....1885.
John Sherman, R., Mansfield..............................1887.

Oregon.
Lafayette Grover, D., Salem ....1883.
John H. Slater, D., La Grande ....1885.

Pennsylvania.
Jas. Donald Cameron, R., Harrisburg......................1885.
John I. Mitchell, R.,....................................1887.

Rhode Island.
Henry B. Anthony, R., Providence .... 1883.
Ambrose E. Burnside, R., Providence .... 1887.

South Carolina.
Manning C. Butler, D., Edgefield C. H. .... 1883.
Wade Hampton, D., Columbia .... 1885.

Tennessee.
Isham G. Harris, D., Memphis .... 1883.
Howell E. Jackson, D., Jackson .... 1887.

Texas.
Richard Coke, D., Waco .... 1883.
Samuel B. Maxey, D., Paris .... 1887.

Vermont.
Justin S. Morrill, R., Strafford .... 1885.
George F. Edmunds, R., Burlington .... 1887.

Virginia.
John W. Johnston, D., Abingdon .... 1883.
William Mahone, Ind. D. .... 1887.

West Virginia.
Henry G. Davis, D., Piedmont .... 1883.
Johnson N. Camden, D., Parkersburg .... 1887.

Wisconsin.
Angus Cameron, R. .... 1885.
Philetus Sawyer, R., Oshkosh .... 1887.

*Resigned their seats May 16, 1881.
Republicans, 37; Democrats, 37; Independent Democrats, 2; Total, 76.

PRINCIPAL OFFICES OF THE SENATE.

President of the Senate—the Vice-President of the United States.

President *pro tempore*.

Chaplain.

Secretary of the Senate.

Chief Clerk.

Principal Executive Clerk.

Principal Legislative Clerk.

Sergeant-at-Arms.
Postmaster.
Superintendent of Folding-Room.
Superintendent of Document-Room.
Five Official Reporters of Debate.

HOUSE OF REPRESENTATIVES.

The present number of Representatives is 293—on the basis of the last apportionment law—one Representative to every 131,425 of population. Congress at its next regular, if not in call session, will make a new apportionment law from the census returns of last year, which will increase the number of Representatives, and the electoral vote for 1884 will probably reach 395. The political complexion of the Forty-seventh Congress, which convenes in December, 1881, will stand: Republicans, 147; Democrats, 137; Nationals, 9.

The term of Congressmen is short, and a complete table to-day would not long be reliable. We give the Representatives of Illinois and Indiana for the Forty-seventh Congress, from March 4, 1881, to March 4, 1883.

ILLINOIS.

First District—Willian Aldrich, R., Chicago.
Second District—George R. Davis, R., Chicago.
Third District—Charles B. Farwell, R., Chicago.
Fourth District—John C. Sherwin, R., Aurora.
Fifth District—Robert M. A. Hawk, R., Mt. Carroll.
Sixth District—Thomas J. Henderson, R., Princeton.
Seventh District—William Cullen, R., Ottawa.
Eighth District—Lewis E. Payson, R., Pontiac.
Ninth District—John H. Lewis, R., Knoxville.
Tenth District—Benjamin F. Marsh, R., Warsaw.

Eleventh District—James W. Singleton, D., Quincy.
Twelfth District—Wm. M. Springer, D., Springfield.
Thirteenth District—Dietrich C. Smith, R., Pekin.
Fourteenth District—Joseph G. Cannon, R., Danville.
Fifteenth District—Samuel W. Moulton, D., Shelbyville.
Sixteenth District—William A. J. Sparks, D., Carlyle.
Seventeenth District—William R. Morrison, D., Waterloo.
Eighteenth District—John R. Thomas, R., Metropolis.
Nineteenth District— Richard W. Townsend, D., Shawneetown.

INDIANA.

First District—William Heilman, R., Evansville.
Second District—Thomas R. Cobb, D., Vincennes.
Third District—Strother M. Stockslager, D., Corydon.
Fourth District—William S. Holman, D., Aurora.
Fifth District—Courtland C. Matson, D., Greencastle.
Sixth District—Thomas M. Browne, R., Winchester.
Seventh District—Stanton J. Peelle, R., Indianapolis.
Eighth District—Robert B. F. Pierce, R., Crawfordsville.
Ninth District—Godlove S. Orth, R., Lafayette.
Tenth District—Mark L. De Motte, R., Valparaiso.
Eleventh District—George W. Steele, R., Marion.
Twelfth District—Walpole G. Colerick, D., Ft. Wayne.
Thirteenth District—William H. Calkins, R., LaPorte.

PRINCIPAL OFFICERS OF THE HOUSE.

Speaker,
Chaplain,
Clerk of the House,

Doorkeeper,
Superintendent of Folding-Room,

Chief Clerk,
Superintendent of Lower Document-Room,
Librarian,
Sergeant-at-Arms,
Superintendent of Upper Document-Room,
Postmaster,
Stenographers—two,
Official Reporters of Debates—five.

OFFICERS OF CONGRESS.

Public Printer—John D. Defrees.
Librarian of Congress—Ainsworth R. Spofford.

THE UNITED STATES.

1. When and by whom settled?
2. When admitted into the Union?
3. Population.
4. Area.
5. Capital.
6. Motto.
7. State elections.
8. Governors take their seats.
9. Term of office.
10. Salary.

ALABAMA.

Alabama was settled by the French in 1713.
Admitted into the Union December 14, 1819.
Population in 1880—1,262,794.
Area—50,722 square miles.
Capital—Montgomery.
Motto—None.
State Election—Tuesday after the first Monday in November.

Governor—Rufus W. Cobb, D.; took his seat on the 28th of November, 1880.

Term of office—two years.

Salary—$3,000.

ARKANSAS

Arkansas was settled by the French in 1670.

Admitted into the Union June 15, 1836.

Population, 1880—802,564.

Area—52,198 square miles.

Capital—Little Rock.

Motto—" Regnant Populi "—The people rule.

State Election—First Monday in November.

It is called the Bear State.

Governor—J. J. Churchill, D.; took his seat January 7, 1881.

Term of office—two years.

Salary—$3,500.

On the 22d of February, 1881, a joint resolution passed the Senate, by a vote of 18 yeas to 5 nays, fixing the pronunciation of the name of this state as "Arkansaw."

CALIFORNIA.

California was settled by the Spaniards in 1769.

Admitted into the Union September 9, 1850.

Population, 1880—864,686.

Area—157,801 square miles.

Capital—Sacramento.

Motto—" Eureka," I have found it.

State Election—The first Tuesday in September.

Governor—George E. Perkins, R.; took his seat December 1, 1879.

Term of office—four years.

Salary—$6,000.

COLORADO.

Colorado was settled by the Americans in 1860.
Admitted into the Union August 1, 1876.
Population in 1880—194,649.
Area—105,000 square miles.
Capital—Denver.
Motto—"Nil Sine Numine"—Nothing can be done without Divine aid.
State Election—Tuesday after the first Monday in November.
Governor—F. W. Pitkin, R.; took his seat January 14, 1881.
Term—two years.
Salary—$3,000.

CONNECTICUT.

Connecticut was settled by the English in 1633.
Admitted into the Union January 9, 1788.
Population in 1880—622,683.
Area—4,750 square miles.
Capital—Hartford.
Motto—"Qui Transtulit Sustinet"—He who brought us over sustains us. Is called the "Nutmeg State."
State Election—First Monday in April.
Governor—H. B. Bigelow, R.; took his seat January 9, 1881.
Term—two years.
Salary—$2,000.

DELAWARE.

Delaware was settled by the Swedes in 1672.
Admitted into the Union December 7, 1787.
Population in 1880—146,654.
Area—2,120 square miles.
Capital—Dover.

Motto—" Liberty and Independence." Called the " Blue Hen State."

State Election—Tuesday after the first Monday in November.

Governor—J. W. Hall, D.; took his seat January 21, 1879.

Term of office—four years.

Salary—$2,000.

FLORIDA.

Florida was settled by the Spaniards in 1564.

Admitted into the Union March 3, 1845.

Population in 1880—267,351.

Area—59,284 square miles.

Capital—Tallahassee.

Motto—" In God is Our Trust."

State Election—The Tuesday after the first Monday in November.

Governor—W. D. Bloxham, D.; took his seat January 1, 1881.

Term of office—four years.

Salary—$3,500.

GEORGIA.

Georgia was settled by the English in 1733.

Admitted into the Union January 2, 1788.

Population in 1880—1,539,083.

Area—58,000 square miles.

Capital—Atlanta.

Motto—" Wisdom, Justice and Moderation."

State Election—The first Tuesday in August.

Governor—A. H. Colquitt, D.; took his seat November 3, 1880.

Term—two years.

Salary—$3,000.

ILLINOIS.

Illinois was settled by the French in 1749.
Admitted into the Union December 3, 1818.
Population in 1880—3,078,636.
Area—55,410 square miles.
Capital—Springfield
Motto—" State Sovereignty, National Union." Called the " Sucker State."
State Election—The Tuesday after the first Monday in November.
Governor—Shelby M. Cullom, R.; took his seat January 8, 1881.
Term of office—four years.
Salary—$6,000.

INDIANA.

Indiana was settled by the French in 1730.
Admitted into the Union December 11, 1816.
Population in 1880—1,978,329.
Area—33,809 square miles.
Capital—Indianapolis.
Motto—None. Is called the " Hoosier State."
State Election—The Tuesday after the first Monday in November.
Governor—Albert G. Porter, R.; took his seat January 8, 1881.
Term of office—four years.
Salary—$5,000.

By a vote of the people at their April election in 1880, Indiana was made a November state, but the decision of a majority of the Supreme Court ruled it not valid, and it "went by the board."

At a special election called by Governor Porter on

the 14th of March, 1881, the people again voted upon the same amendment; it was carried and now is law.

Among the bills passed at the call session in March, 1881, by the Indiana Legislature, were the prohibitory and female suffrage amendments to the constitution; the former passed the House by the vote, ayes 55, noes 35; passed the Senate, ayes 26, noes 20.

The female suffrage bill was adopted in the House by a vote yeas 62, nays 24; in the Senate, yeas 27, nays 18. These proposed amendments to become a part of the organic law will have to receive a majority of the votes of both branches of the next Legislature, which meets in January, 1883, then be submitted by an election to the people, which will require a majority of the votes cast upon each amendment separately. An effort will no doubt be made by the friends of these measures to engraft them upon some of the party platforms, and members to the next Legislature may be elected or defeated by favoring the enactment of these measures or opposing their adoption.

The school law remains as published in 1877, save one change, that women are now "*eligible to any office under the general or special school laws of the State of Indiana.*"

IOWA.

Iowa was settled by the Americans in 1835.

Admitted into the Union December 28, 1846.

Population in 1880—1,624,620.

Area—55,045 square miles.

Capital—Des Moines.

Motto—" Our Liberties we Prize, Our Rights we will Maintain." It is known as the " Hawkeye State."

State Election—Second Tuesday in October.

Governor—John H. Gear, R.; took his seat January 4, 1880.
Term—two years.
Salary—$3,000.

KANSAS.

Kansas was settled by the Americans in 1850.
Admitted into the Union January 29, 1861.
Population in 1880—995,966.
Area—81,318 square miles.
Capital—Topeka.
Motto—"Ad astra per aspera."—To the stars through difficulties.
State Election—Tuesday after the first Monday in November.
Governor—J. P. St. John, R.; took his seat January 13, 1881.
Term—two years.
Salary—$3,000.

KENTUCKY.

Kentucky was settled by Virginians in 1775.
Admitted into the Union June 1, 1792.
Population in 1880—1,648,699.
Area—37,630 square miles.
Capital—Frankfort.
Motto—"United We Stand, Divided We Fall." Called the "Blue Grass State."
State Election—The first Monday in August.
Governor—L. P. Blackburn, D.; took his seat September 2, 1879.
Term—four years.
Salary—$5,000.

LOUISIANA.

Louisiana was settled by the French in 1699.
Admitted into the Union April 12, 1812.
Population in 1880—940,103.
Area—41,346 square miles.
Capital—Baton Rouge.
Motto—" Union and Confidence." It is known as the " Creole State."
State Election—The first Monday in November.
Governor—L. A. Wiltz, D.; took his seat January 10, 1880.
Term of office—four years.
Salary—$4,000.

MAINE.

Maine was settled by the English in 1630.
Admitted into the Union March 3, 1820.
Population in 1880—648,945.
Area—35,000 square miles.
Capital—Augusta.
Motto—" Dirigo "—I direct. Called the " Pine Tree State."
State Election—The second Monday in September.
Governor—H. M. Plaisted, N.; took his seat January 1, 1881.
Term—two years.
Salary—$2,000.

MARYLAND.

Maryland was settled by the English in 1634.
Admitted into the Union April 28, 1788.
Population in 1880—934,632.
Area—11,124 square miles.
Capital—Annapolis.

Motto—" Crescite et Multiplicamine "—Increase and Multiply.

State Election—Tuesday after the first Monday in November.

Governor—W. T. Hamilton, D.; took his seat January 1, 1880.

Term—four years.

Salary—$4,500.

MASSACHUSETTS.

Massachusetts was settled by the English in 1620.

Admitted into the Union February 6, 1788.

Population in 1880—1,783,086.

Area—7,800 square miles.

Capital—Boston.

Motto—" Ense petit placidam sub libertate quietem" —By the Sword she seeks placid rest in Liberty. The " Bay State."

State Election—Tuesday after the first Monday in January.

Governor—John D. Long, R.; took his seat January 7, 1881.

Term—one year.

Salary—$5,000.

MICHIGAN.

Michigan was settled by the French in 1670.

Admitted into the Union January 26, 1837.

Population in 1880—1,636,331.

Area—56,451 square miles.

Capital—Lansing.

Motto—" Tuebor"—I will defend; " Si quœris peninsulam amœnam circumspice"—If you seek a pleasant peninsula, look around you.

State Election—Tuesday after the first Monday in November.

Governor—D. H. Jerome, R.; took his seat January 1, 1881.

Term—two years.

Salary—$1,000.

MINNESOTA.

Minnesota was settled by Americans in 1847.

Admitted into the Union May 4, 1858.

Population in 1880—780,807.

Area—83,531 square miles.

Capital—St. Paul.

Motto—" L'Etoile du Nord "—The Star of the North.

State Election—Tuesday after the first Monday in January.

Governor—John H. Pillsbury, R.; took his seat January 6, 1880.

Term—two years.

Salary—$3,800.

MISSISSIPPI.

Mississippi was settled by the French in 1716

Admitted into the Union December 10, 1817.

Population in 1880—1,131,592.

Area—47,156 square miles.

Capital—Jackson.

Motto—None.

State Election—Tuesday after the first Monday in November.

Governor—John M. Stone, D.; took his seat January 7, 1878.

Term—four years.

Salary—$4,000.

MISSOURI.

Missouri was settled by the French in 1763.
Admitted into the Union March 2, 1821.
Population in 1880—2,168,804.
Area—65,350 square miles.
Capital—Jefferson City.
Motto—" Salus populi suprema lex esto "—Let the welfare of the people be the supreme law.
State Election—Tuesday after the first Monday in November.
Governor—T. J. Crittenden, D.; took his seat January 8, 1881.
Term—four years.
Salary—$5,000.

NEBRASKA.

Nebraska was settled by the Americans in 1850.
Admitted into the Union March 1, 1867.
Population in 1880—452,432.
Area—75,995 square miles.
Capital—Lincoln.
Motto—" Equality before the Law."
State Election—Second Tuesday in October.
Governor—Albinus Nance, R.; took his seat January 7, 1881.
Term—two years.
Salary—$2,500.

NEVADA.

Nevada was settled by Americans in 1860.
Admitted into the Union March 21, 1864.
Population in 1880—62,265.
Area—112,090 square miles.
Capital—Carson City.

Motto—" Volens et Potens "—Willing and powerful.
State Election—The first Tuesday in November.
Governor—John H. Kinkhead, R.; took his seat January 6, 1879.
Term—four years.
Salary—$6,000.

NEW HAMPSHIRE.

New Hampshire was settled by the English in 1623.
Admitted into the Union June 21, 1788.
Population in 1880—346,984.
Area—9,289 square miles.
Capital—Concord.
Motto—None. Is known as the " Old Granite State."
State Election—Second Tuesday in March.
Governor—Charles H. Bell, R.; took his seat June 1, 1881.
Term—two years.
Salary—$1,000.

NEW JERSEY.

New Jersey was settled by the Swedes in 1627.
Admitted into the Union December 18, 1787.
Population in 1880—1,130,983.
Area—8,320 square miles.
Capital—Trenton.
Motto—" Liberty and Independence."
State Election—Tuesday after the first Monday in November.
Governor—George C. Ludlow, D.; took his seat January 18, 1881.
Term—Three years.
Salary—$5,000.

NEW YORK.

New York was settled by the Dutch in the year 1613.
Admitted into the Union July 26, 1788.
Population in 1880—5,083,810.
Area—47,000 square miles.
Capital—Albany.
Motto—" Excelsior," higher. Called the " Empire State."
State Election—Tuesday after the first Monday in November.
Governor—Alonzo B. Cornell, R.; took his seat January 1, 1880.
Term—three years.
Salary—$10,000.

NORTH CAROLINA.

North Carolina was settled by the English in 1650.
Admitted into the Union November 21, 1789.
Population in 1880—1,400,000.
Area—50,704 square miles.
Capital—Raleigh.
Motto—None. Sometimes called the " Turpentine State."
State Election—First Thursday in August.
Governor—Thos. J. Jarvis, D.; took his seat January 1, 1881.
Term—four years.
Salary—$3,000.

OHIO.

Ohio was first settled by Virginians in 1788.
Admitted into the Union April 30, 1802.
Population in 1880—3,198,239.
Area—39,964 square miles.
Capital—Columbus.

Motto—"Imperium in Imperio"—An empire in an empire.

State Election—Second Tuesday in October.

Governor—Charles Foster, R.; took his seat January 12, 1880.

Term—two years.

Salary—$4,000.

OREGON.

Oregon was settled by the English in 1796.

Admitted into the Union February 14, 1856.

Population in 1880—174,767.

Area—95,274 square miles.

Capital—Salem.

Motto—"Alis volat propriis"—She flies with her own wings.

State Election—First Monday in June.

Governor—William W. Thayer, D.; took his seat September 11, 1878.

Term—four years.

Salary—$4,500.

PENNSYLVANIA.

Pennsylvania was settled by the English in 1682.

Admitted into the Union December 12, 1787.

Population in 1880—4,282,786.

Area—46,000 square miles.

Capital—Harrisburg.

Motto—"Virtue, Liberty and Independence." Called the "Keystone State."

State Election—Second Tuesday in October.

Governor—H. M. Hoyt, R.; took his seat January 21, 1879.

Term—four years.

Salary—$10,000.

RHODE ISLAND.

Rhode Island was first settled by the English in 1631.
Admitted into the Union May 29, 1790.
Population in 1880—276,528.
Area—1,306 square miles.
Capitals—Newport and Providence.
Motto—" Hope."
State Election—First Wednesday in April.
Governor—A. H. Littlefield, R.; took his seat May 27, 1881.
Term—one year.
Salary—$1,000.

SOUTH CAROLINA.

South Carolina was settled by the English in 1689.
Admitted into the Union May 23, 1788.
Population in 1880—995,706.
Area—34,000 square miles.
Capital—Columbia.
Motto—"Animis opibusque parati "—Ready in will and deed.
State Election—Fourth Monday in November.
Governor—Johnson Hagood, D.; took his seat December 30, 1880.
Term—two years.
Salary—$3,500.

TENNESSEE.

Tennessee was settled by the Virginians in 1765.
Admitted into the Union June 1, 1796.
Population in 1880—1,542,463.
Area—45,600 square miles.
Capital—Nashville.

Motto—"Agriculture and Commerce." Known as the Big Bend State."

State Election—First Monday in August.

Governor—Alvin Hawkins, R.; took his seat January 15, 1881.

Term—two years.

Salary—$4,000.

TEXAS.

Texas was settled by the Spaniards in 1690.

Admitted into the Union March 1. 1845.

Population in 1880—1,592,509

Area—274,356 square miles.

Capital—Austin.

Motto—None. Is known as the "Lone Star State."

State Election—Tuesday after the first Monday in November.

Governor—O. M. Roberts, D.; took his seat January 16, 1881.

Term—two years.

Salary—$4,000.

VERMONT.

Vermont was settled by the English in 1763.

Admitted into the Union February 18, 1791.

Population in 1880—332,286.

Area—9,612 square miles.

Capital—Montpelier.

Motto—"Freedom and Unity." Called the "Green Mountain State."

State Election—First Tuesday in September.

Governor—Roswell Farnham, R.; took his seat October 4, 1880.

Term—two years.

Salary—$1,000.

VIRGINIA.

Virginia was settled by the English in 1607
Admitted into the Union June 26, 1788.
Population in 1880—1,512,803.
Area—38,352 square miles.
Capital—Richmond.
Motto—" Sic Semper Tyrannis "—Thus always with Tyrants.
State Election—Tuesday after the first Monday in November.
Governor—F. W. Holliday, D.; took his seat January 1, 1878.
Term—four years.
Salary—$5,000.

WEST VIRGINIA.

West Virginia being a part of Virginia, in 1607 was settled by the English. It was separated from Virginia proper, and admitted as a state December 31, 1862.
Population in 1880—618,443.
Area—23,000 square miles.
Capital—Wheeling.
Motto—Montani Semper Liberi "—Mountaineers are always Free.
State Election—Fourth Thursday in October.
Governor—J. B. Jackson, D.; took his seat March 4, 188
Term—four years.
Salary—$2,700.

WISCONSIN.

Wisconsin was settled by the Americans in 1831
Admitted into the Union March 3. 1847.
Population in 1880—1,315,486.
Area—53,924 square miles.
Capital—Madison.

Motto—"The civilized man succeeds the barbarous."
The "Badger State."

State Election—Tuesday after the first Monday in November.

Governor—William E. Smith, R.; took his seat January 5, 1880.

Term—two years.

Salary—$5,000.

Republican Governors, 20; Democratic Governors, 17; National, 1.

To find the number of acres of any state, multiply its area by 640.

TERRITORIES.

The Governors of Arizona, Dakota, Idaho, Montana, New Mexico, Utah, Washington and Wyoming serve each four years and receive each $2,600 per annum.

The Alaska and Indian Territories have not yet been organized.

TERRITORIES	Organized.	No. Co.	Population.	
			1870.	1880.
Arizona	1863	7	9,658	40,441
Dakota	1864	36	14,181	134,502
Idaho	1863	12	15,000	32,611
Montana	1864	10	20,595	39,157
New Mexico	850	12	91,874	118,430
Utah	50	23	86,786	143,907
Washington	53	23	23,955	75,120
Wyoming	868	5	9,118	20,788

Alaska was ceded to the United States, March 30, 1867, for the sum of $7,200,000.

UNITED STATES MONEY.

Gold, silver, copper, nickel, national bank notes, legal tender notes or greenbacks, and the silver certificates constitute our money.

GOLD:—One dollar piece, two and a half, or quarter eagle; three dollar piece; five dollar, or half-eagle; ten dollar, or eagle; twenty dollar, or double-eagle.

SILVER:—Three, five, ten, twenty, twenty-five, and fifty-cent pieces; standard dollar of $412\frac{1}{2}$ grains, and the trade dollar of 420 grains. The first silver dollars coined weighed 416 grains, standard silver.

COPPER:—One and two-cent pieces.

NICKEL:—Three and five-cent pieces.

NATIONAL BANK NOTES are issued in denominations of ones, twos, fives, tens, twenties, fifties, one hundred, five hundred, and one thousand-dollar notes. The one and two-dollar notes were discontinued November 1, 1878, in view of the provision :—Section 5,175 of the Revised Statutes, "that not more than one-sixth part of the notes furnished to any association shall be of a less denomination than five dollars, and that after specie payments are resumed no association shall be furnished with notes of a less denomination than five dollars.

THE LEGAL TENDER or GREENBACK notes are of the same denomination as the national bank notes, with the addition of the five and ten thousand dollar notes.

SILVER CERTIFICATES:—Tens, twenties, fifties, one hundred, and five hundred dollar notes.

Our coined gold and silver is nine parts fine and one part alloy. Silver is alloyed with copper; gold with copper and silver.

In testing the purity of gold by analysis at the mints, weight is not taken into consideration. Any amount, be it large or small, is called an "assay pound," which is taken up and analyzed, and its fineness is fixed according to the proportionate amount of purity to the amount of impurity. If a mass of it is found to contain 12 carats gold, it is then 12-24ths pure, and 12-24ths or $\frac{1}{2}$ impure. If it is 16 carats gold, then it is 16-24ths pure, and 8-24ths impure; if 18 carats gold, it is 18-24ths or $\frac{3}{4}$ pure gold and 6-24ths or $\frac{1}{4}$ impure; if 24 carats gold, it is entirely pure gold, free from all alloy. The table used:

4 quarters...............	1 assay grain.
4 grains..................	1 carat.
24 carats	1 assay pound.

VALUABLE COINS.

Some Numismatic Information of Interest.

Of all the decimal United States coins the most valuable is the silver dollar of 1804, which is excessively rare. Specimens are worth from $500 to $1,000 each, according to the nearness with which they approach perfection. The coinage of this year was very limited, and there were no more dollars coined until 1836. "Proofs" of the last-named year are worth $10, and good examples $5. There was nothing done in dollars in 1837, and the issues of 1838 and 1839 are rare enough to raise the quotations for good specimens to $40 each. From that date forward to 1873, when the trade dollar came in, there is no break in the line of dollars, but from 1850 to 1856, inclusive, they are quoted as "rare" or "scarce" those of 1851 and 1852 being worth $35 or $40 each. Previous to 1804 the value of a good specimen varies from $1.75 for 1799, to $5 for 1798, and $4 for 1801, save that the first date of all (1794), which is very rare, brings $50. Some of the early dates are made peculiarly valuable by reason of variation in the number and style of stars, etc., there being three variations of 1798 and five of 1795.

Of the silver half-dollars, those of 1796 and and 1797 are the most valuable, choice samples of these dates being worth from $15 to $20. Good ones of other years previous to 1806 will bring from $2 to $4. One of this class of 1815 is quoted at $2.50, and then they are of little rarity until 1836, when a specimen with reeded edge and head of 1837 is valued at $3 or $4. The other issue of this year is worth $1. The next dates of note are 1850, 1851 and 1852, valued at $1.50, $2.50 and $3 respectively. More recent dates are only valuable to collectors when in perfect conditions, "proofs" of later issues only being desired, and they range in worth from $1.25 to $8.

Quarter dollars are likewise a speculative issue, and, therefore, favorites with dealers, particularly the dates 1823 and 1827, which are excessively rare, and command from $45 to $75 each. The 1853 issue, without arrows, is also much sought after, fair specimens bringing from $6 to $8. The only other dates worth over $1 for "good" examples are: 1824, $1.50; 1822, $2; 1819, $1.75; 1815, $2; 1807, $2; 1806, $2; 1805, $1.50; 1804, $4 and 1796, $4.

Silver dimes are still more valuable as a class than the quarters, their smaller size and more general circulation having made good specimens rather scarce in all the earlier dates. From 1828 back to 1796 they range in worth from $1 to $7, except in five instances. The high rates are: 1824, $2.50; 1825, $5; 1811, $2.50; 1809, $3: 1807, $2; 1803, $3; 1802, $6; 1801, $5;

1800, $7; 1798, $5; 1797, $5 and 1796, $3. And 1840, with a draped figure of Liberty, like 1841, is worth $1, as is a good issue of 1846.

Of all the minor coins, however, an 1802 half-dime is the chief in cost, the price ranging from $75 to $200, according to quality. A good specimen of many other dates is, nevertheless, a handy thing to have, as will be noted by the following quotations: 1794, $4; 1796, $4; 1797, $2; 1800, $1.25; 1801, $6; 1803, $4; 1804, $4; 1840 (with drapery), $1; 1846, $1.75. From that date until 1873, when the coinage closed, no unusual worth attaches to this class. A first-class specimen of the last-named date is worth 50 cents, however.

For the three-cent silver piece there is but one little call, as their period only reaches from 1851 to 1873, including both these years. By far the most valuable of all of them is the 1855, a perfect specimen of which is worth $2. From 1863 to 1869 an uncirculated one is worth 58 cents. All the other dates are of small value.—*Ind. Sentinel.*

EMIGRATION TO THE UNITED STATES, BY COUNTRIES, DURING 60 CALENDAR YEARS—1829-1879.

[From the American Almanac for 1881.]

GREAT BRITAIN.				SUMMARY	
England	894,444	Austria-Hung'y	65,588		
		Belgium	23,267		
Ireland	3,065,761	Denmark	48,620	Europe	8,746,921
		France	313,716	Asia	228,047
		Germany	3,002,027		
Scotland	159,547	Greece	385	Africa	1,631
		Italy	70,181	Bri. America	568,941
Wales	17,893	Netherlands	44,319	Other American countries	97,007
		Poland	14,831		
		Portugal	9,062		
Great Britain, Not specified	560,453	Russia	38,316		
		Spain	28,091	Pacific	10,474
		Sweden & Norway	306,092	All other	255,778
		Switzerland	83,709		
Total from British Isles	4,698,098	Turkey	619	Grand aggregate	9,908,799
		Total from Europe	8,746,921		

CHINESE EMIGRATION INTO THE UNITED STATES FOR EACH CALENDAR YEAR FROM 1855 TO 1880 INCLUSIVE.

Year.	No.	Year.	No.	Year.	No.	Year.	No.
1855	3,526	1862	3,633	1869	14,902	1876	16,879
1856	4,733	1863	7,214	1870	11,943	1877	10,379
1857	5,944	1864	2,795	1871	6,039	1878	8,468
1858	5,128	1865	2,942	1872	10,642	1879	9,189
1859	3,457	1866	2,385	1873	18,154	1880 Jan. to Ju'e.	4,018
1860	5,467	1867	3,863	1874	16,651		
1861	7,518	1868	10,684	1875	19,033	Total	215,586

NOTE.—The statement is made that nearly one-half of all the Chinese who have arrived in the United States have returned to their native country.

PAY OF THE ARMY OF THE UNITED STATES.

[From the American Almanac for 1881.]

GRADE.	Pay of Officers in Active Service. Yearly Pay.			Pay of Retired Officers. Yearly Pay.		
	First 5 years' service.	After 5 years' service.	After 10 yrs. service.	First 5 years' service.	After 5 years' service.	After 10 yrs. service.
		10 p. c.	20 p. c.			
General	$13,500					
Lieutenant-General	11,000					
Major-General	7,500			$5,625		
Brigadier-General	5,500			4,125		
Colonel	3,500	$3,850	$4,200	2,625	$2,887	$3,150
Lieutenant-Colonel	3,000	3,300	3,600	2,250	2,475	2,700
Major	2,500	2,750	3,000	1,875	2,062	2,250
Captain, mounted	2,000	2,200	2,400	1,500	1,650	1,800
Captain, not mounted	1,800	1,980	2,160	1,350	1,485	1,620
Regimental Adjutant	1,800	1,980	2,160			
Regimental Quarterm'r	1,800	1,980	2,160			
1st Lieutenant, mounted	1,600	1,760	1,920	1,200	1,320	1,440
1st Lieutenant, not m't'd	1,500	1,650	1,800	1,125	1,237	1,350
2d Lieutenant, mounted	1,500	1,650	1,800	1,125	1,237	1,350
2d Lieutenant, not m't'd	1,400	1,540	1,680	1,050	1,155	1,260
Chaplain	1,500	1,650	1,800	1,350	1,485	1,620

QUARTERS, FUEL, AND FORAGE ALLOWED TO ARMY OFFICERS.

By act of June 18, 1878, all allowance or commutation for fuel was prohibited, but wood is furnished at $3 per cord, out of the pay of officers. Forage is furnished only in kind, and only to officers actually in the field or west of the Mississippi, on the basis of five horses for the General of the Army, four for the Lieutenant-General, three each for a major or brigadier-general, and two each for a colonel, lieutenant-colonel, major, mounted captain or lieutenant, adjutant and regimental quartermaster. Quarters are furnished on the following basis: General (commutation for quarters), $125 per month; Lieutenant-General, $70 per month; major-general, six rooms; brigadier-general or colonel, five rooms; lieutenant-colonel or major, four rooms; captain or chaplain, three rooms; and first or second-lieutenant, two rooms—all of which may be commuted at $10 per room per month.

<small>Note.—The law provides that no allowance shall be made to officers in addition to their pay, except quarters and forage furnished in kind.

Mileage at the rate of eight cents per mile is allowed for travel under orders.

The pay of cadets at the U. S. Military Academy, West Point, was placed at $540 per annum, by act of August 7, 1876, instead of $500 and one ration per diem (equivalent to $609.50), by former laws.

The pay of privates runs from $156 ($13 a month and rations) for first two years, to $21 a month after twenty years' service.</small>

MISCELLANEOUS.

There are 850 daily and 7,500 weekly newspapers published in the United States say nothing of the periodicals and various other kinds of publications. We here append seven leading newspaper states:

New York....115 dailies.	New York...800 weekly.
Pennsylvania . 90 "	Illinois......650 "
Illinois....... 70 "	Pennsylvania..620 "
Ohio......... 50 "	Ohio.........525 "
California..... 48 "	Iowa.........450 "
Missouri...... 40 "	Missouri......375 "
Indiana....... 38 "	Indiana.......350 "

Number of newspapers and periodicals published in seven of the largest cities in the United States:

Rank by Population, 1880.	Rank by the No. of Periodicals.
New York......1,206,590.	New York....435 per'd's.
Philadelphia.... 846,984.	Chicago.......165 "
Brooklyn...... 566,689.	Philadelphia...155 "
Chicago........ 503,304.	Boston........115 "
Boston......... 362,535.	St. Louis...... 75 "
St. Louis....... 350,522.	Baltimore..... 45 "
Baltimore...... 332,190.	Brooklyn..... 20 "

SURVEYOR'S MEASURE.

7.92 inches	1 link.
100 links	1 chain.
80 chains	1 mile.

7.92 multiplied by 100 equal 792 inches, divided by 12 equal 66 feet, or 1 chain; and 66 feet divided by 3 feet equal 22 yards, divided by 5½ yards equal 4 rods, or 1 chain; hence, the chain used by surveyors is 4 rods or 66 feet long. If you wish to lay off a square acre, measure 209 feet on each side, and you have it within one inch, nearly.

BARREL MEASURE.

196	lbs. flour	one barrel.
200	lbs. pork	one barrel.
200	lbs. beef	one barrel.
600	lbs. rice	one barrel.
25	lbs. powder	one barrel.
280	lbs. salt	one barrel.
100	lbs. nails	one keg.
56	lbs. butter	one firkin.
84	lbs. butter	one tub.

TO MEASURE CORN IN A CRIB, BIN OR WAGON.

First reduce the dimensions all to the same denomination; then multiply the length, depth and width together; if the dimensions were in feet, you will have cubic feet, multiply by 1728, the number of cubic inches in one cubic foot, this will reduce the dimensions to cubic inches; divide this product by 2150.42, the number of cubic inches in a bushel; the quotient thus obtained will be the answer sought; but if part of the dimensions be in feet and part in inches, first reduce all to inches, then multiply the three dimensions together and divide the product by 2150.42, and the quotient will represent the number of bushels required. The above rule applies to wheat, oats, shelled corn, etc. To find the number of bushels of corn in the ear, use the above rule in reducing your cribs, bins, etc., to cubic feet or cubic inches as the case may require, but divide by .3888 instead of 2150 42, and you obtain the number of bushels sought.

One bushel of dry measure contains 2150.42 cubic inches.

One gallon, dry measure contains 268.8 cubic inches.
One gallon, wine, contains 231 cubic inches.
One gallon, beer or ale, contains 282 cubic inches.

EXCHANGE.

Bills of Exchange, Drafts, or Checks are used to pay debts at a distance without having to transmit actual money. There are two kinds of Exchange, *Foreign and Domestic.*

The person or bank on whom the draft is drawn is called the *drawee*; and the person to whose order it is to be paid is called the *payee*.

COMMON FORM OF A DOMESTIC DRAFT.

$1,000.   Burlington, Kansas, June 1, 1881.

Pay to the order of A. H. Smith, one thousand dollars and charge

To Kramer & Son, Bankers, New York.

FORM OF A FOREIGN DRAFT.

Exchange £1000.   Boston, July 4, 1881.

At ninety days' sight of this first exchange (the second and third of the same date and tenor unpaid), pay to the order of Thomas J. Brant, one thousand pounds, without further notice.

To William Tell, Merchant, London.   Smith & Brown.

A promissory note is a written agreement by one party to pay a certain sum of money to another party at a specified time.

COMMON FORM OF A NOTE.

$100.   Terre Haute, Ind., July 1, 1881.

Twelve months after date. I promise to pay to the order of John Roy, one hundred dollars, VALUE RECEIVED, without any relief whatever from valuation or appraisement laws, with five per cent. interest from date until paid. Samuel Thomas.

The words "VALUE RECEIVED" are always essential in plain notes of hand, but immaterial to a Bill of Exchange.

WEIGHTS OF WOOD.

Wood	lbs. in one cubic foot
Pine	23
Poplar	26
Dry Cedar	28
Dry Willow	30
Dry Chestnut	33
Cedar	35
Dry Elm	36
Dry Sycamore	36
Dry Walnut	38
Green Willow	38
Dry Oak	39
Green Sycamore	40
Dry Beach	43
Maple	49
Dry Ash	52
Dry Mahogany	53
Green Chestnut	54
Green Walnut	57
Green Elm	58
Green Oak	68

One cord of dry hickory weighs 4,400 pounds; one cord of dry maple, 2,600 pounds.

Our best charcoal is obtained from oak, chestnut, maple and beach woods.

Our mineral coal we obtain from beds in the earth; these beds vary from one inch to forty feet in thickness; the bed at Wilkesbarre, Pa., is 37 feet, while at Pittsburg averages 8 feet. From Pottsville, Lehigh, and

Wilkesbarre, we get the anthracite coal; from Pittsburg the bituminous coal. The use of coal has become universal in all kinds of manufactories throughout the world. It is said that two ounces of coal properly arranged will evaporate one pint of water; this will produce 216 gallons of steam, which is capable of raising 37 tons 12 inches high.

Names of principal rivers arranged in the order of length by English statute miles:

River	Continent	Length
Mississippi	North America	3,700 miles.
Amazon	South America	3,540 "
Yenesei	Asia	3,320 "
Yang-tse-Kiang	Asia	3,310 "
Missouri	North America	3,100 "
Volga	Europe	2,760 "
Lena	Asia	2,760 "
Amoor	Asia	2,730 "
Obi	Asia	2,670 "
Hoang Ho	Asia	2,620 "
Nile	Africa	2,570 "
Indus	Asia	2,250 "
Rio de la Plata	South America	2,200 "
St. Lawrence	North America	2,070 "
Rio Grande	North America	2,000 "
Arkansas	North America	2,000 "
Ganges	Asia	1,930 "
Danube	Europe	1,720 "
Euphrates	Asia	1,710 "
Elbe	Europe	780 "
Rhine	Europe	690 "

The velocity of rivers vary; some flow at the rate of two or three feet per second, while others move five or six feet per second. The mean velocity of the Mississippi at its mouth is 2.95 or nearly three feet per second. In large, deep rivers it is said that three inches of declivity to the mile, will give a velocity of three miles per hour to the water. The Mississippi has its tributaries in the heights of the Rocky Mountain chain of the West and the Appalachian chain of the East and South, all flowing to one great basin that forms the waters of the Mississippi; like causes produce the great Amazon of South America.

The Mississippi is said to empty annually, on an average, 19,500,000,000,000 of cubic feet of water into the ocean, and 812,500,000,000 pounds of silt into the Gulf.

The depth of the oceans vary from 10,000 to 50,000 feet; from Newfoundland to Ireland it is said to range from 10,000 to 15,000 feet, while further south it is much deeper. The true basin of the ocean is where the deep water begins; from the coast of New Jersey the slope runs out, it is said, 80 miles before striking the basin, with only one foot slope in every 700 feet of the distance.

As we can know the purity of water, it has been taken as a standard for comparing the weights of all bodies in solid or liquid form.

To find how much heavier or lighter a given substance is than water:—*Divide the weight of the given bulk of substance by the weight of an equal bulk of water, and the quotient will be the specific gravity required.*

The specific gravity of the following solids and liquids, with pure water as the standard, are:

Distilled water	1.000.	Copper	8.870.
Platinum	21.500.	Iron	7.800.
Gold	19.360.	Marble	2.830

Mercury .... 13.600.   Anthracite coal.... 1.800.
Lead .... 11.450.   Alcohol........... 0.800.
Silver.......... 10.500.   Ether .... 0.720.

A cubic foot of water weighs 62½ pounds avoirdupois, or exactly 1,000 ounces; hence, the specific gravity of gold, 19.360 multiplied by 1,000 ounces, equal 19,360, the number of ounces in one cubic foot of gold; so the number of ounces in a cubic foot of any of the solids or liquids may be obtained by multiplying the specific gravity by 1,000, the standard of comparison.

It is estimated that a wire of lead one-twelfth of an inch in diameter, will sustain 27 pounds; of tin, 34 pounds; of zinc, 109 pounds; of gold, 150 pounds; of silver, 187 pounds; of platinum, 274 pounds; of copper, 300 pounds; of iron, 545 pounds; and that cords of the same diameter, of flax, 1,175 pounds; of hemp, 1,600 pounds; of silk, 3,400.

HEIGHT OF MOUNTAIN.
Asia.
Mount Everest, Himalaya Mountains, India, 29,000 feet.
" Elburz, Caucasian " Russia, 18,493 "

Africa.
Mount Killamand Jara, Mountain of Moon, 20,000 feet.
" Abba Yared, Abyssinia .... 15,200 "

Europe.
Mont Blanc, the Alps, Sardinia .... 15,750 feet.
" Parnassus, Greece................ 8,000 "

South America.
Mount Tupungata, the Andes, Chili ...... 22,450 feet.
" Cotopaxi, Ecuador............... 18,875 "

North America.

Mount St. Elias, Coast Range, Russian Pos. 17,900 feet.
" Hooker, Rocky Mts., British A..... 15,700 "
Long's Peak, " " U. S. A....... 13,575 "

This table gives the highest peaks of mountains in the countries of the world; while the mean height is much less. Dr. Dana, in his work on geology, gives the following figures as estimated: "Of Europe, 670 feet; Asia, 1,150 feet; North America, 748 feet; South America, 1,132 feet; Africa, probably, 1,600 feet; and further, that if the material in the Pyrenees was spread equally over Europe it would raise the surface only 6 feet; and that of the Alps only 22 feet."

GRAIN AND PRODUCE TABLE.

Legal number of pounds to the bushel of articles in the following States:

ARTICLES.	Ind.	Ill.	Iowa	Mo.	Wis.	Minn	Kan.
Wheat	60	60	60	60	60	60	60
Corn, shelled	56	56	56	56	56	56	56
Corn, in ear	68	70	70	70	70	70	70
Oats	32	32	33	35	32	32	32
Barley	48	48	48	48	48	48	50
Rye	56	56	56	56	56	56	56
Buckwheat	50	52	52	52	40	52	50
Broom Corn Seed	30	30	30	30	30	30	..
White Beans	60	60	60	60	60	60	60
Castor Beans	46	46	46	46	46	46	46
Irish Potatoes	60	60	60	60	60	60	60
Sweet Potatoes	55	55	46	55	55	50	55
Turnips	55	55	55	55	55	57	55
Onions	57	57	57	57	57	57	57
Peas	60	60	60	60	60	60	..
Dried Peaches	33	28	33	33	28	28	33
Dried Apples	25	54	24	24	28	28	24
Corn Meal	50	48	48	48	48	50	50
Bran	20	20	20	20	20	20	20
Malt	38	38	38	38	38	38	32
Hungarian Grass Seed	48	48	45	48	48	45	55
Hemp Seed	44	44	44	44	44	44	44
Flax Seed	56	56	56	56	56	56	56
Stone Coal	70	80	80	..	..	..	80
Charcoal	22	22	22	22	22	22	..
Unslacked Lime	80	80	80	80	80	80	80
Coarse Salt	50	50	50	50	50	50	..
Plastering Hair	8	8	8	8	8	8	8
Clover Seed	60	60	60	60	60	60	60
Timothy Seed	45	45	45	..	46	45	45
Red Top Seed	14	14	14	14	14	14	..
Blue Grass Seed	14	14	14	14	14	14	14

[From the American Almanac for 1881.]

RELIGIOUS DIVISIONS OF THE WORLD.

[Estimates from Schem's Statistics of the World.]

CHRISTIANS—viz.: { Roman Catholics......201,000,000
Protestants..........106,000,000 } 388,000,000
Eastern Churches ....81,000,000 }

Buddhists..........340,000,000 | Followers of Confucius.80,000,000
Mohammedans......201,000,000 | Sinto Religion..........14,000,000
Brahminism........175,000,000 | Judaism 7,000,000

	Whole Population.	Roman Catholic.	Protestants.	Eastern Churches.
America	84,500,000	47,200,090	30,000,000	
Europe	301,600,000	147,300,000	71,800,000	69,350,000
Asia	798,000,000	4,700,000	1,800,000	8,500,000
Africa	203,300,000	1,100,000	1,200,000	3,200,000
Australia & Polynesia	4,400,000	400,000	1,500,000	
Total	1,392,000,000	201,200,000	106,300,000	81,050,000

POPULATION OF THE UNITED STATES FROM 1790 TO 1880.

1790	3,929,241	1850	23,191,876
1800	5,308,483	1860	31,443,321
1810	7,239,881	1870	38,558,371
1810	9,633,822	1880	50,152,554
1830	12,866,020	1890	
1840	17,069,453	1900	

SCHEDULE OF UNITED STATES PATENT FEES.

On filing each application for a Patent	$15
On issuing each Original Patent (17 years)	20
On application for Re-issue	30
On application for extension	50
On granting every extension of Patent (7 years)	50
On each Caveat	10
On appeal to Examiners-in-Chief	10
On appeal to Commissioner of Patents	20
On filing a Disclaimer	10
On application for Design (3½ years)	10
On application for Design (7 years)	15
On application for Design (14 years)	30
On each Trade-Mark (30 years)	25
On each Label (28 years)	6

Note.—By decision of the Supreme Court of the United States, rendered Nov. 17, 1879, the Trade-mark law of July 8, 1870, by which Trade-marks were for the first time recognized and protected by act of Congress, was declared unconstitutional. The registry of Trade-marks at the Patent Office is, however, continued to such as seek the benefit of a record, without regard to the ultimate validity of the right.

TABLE COMPARING THE POPULATION OF FIFTY CITIES FOR 1870 AND 1880.

City	1870	1880
New York, N. Y	942,292	1,206,590
Philadelphia, Pa	674,022	846,984
Brooklyn, N. Y	396,009	566,689
Chicago, Ill	298,977	503,304
Boston, Mass	250,526	362,535
St. Louis, Mo	310.864	350,522
Baltimore, Md	267,354	332,190
Cincinnati, Ohio	216,239	255,708
San Francisco, Cal	149,743	233,956
New Orleans, La	191,418	216,140
Cleveland, Ohio	92,829	160,142
Pittsburg, Pa	86,076	156,381
Buffalo, N. Y	117,714	155,137
Washington, D. C	109,199	147,307
Newark, N. J	105,059	136,400
Louisville, Ky	100,753	123,645
Jersey City, N. J	82,546	120,728
Detroit, Mich	79,577	116,342
Milwaukee, Wis	71,440	115,578
Providence, R. I	68,904	104,850
Albany, N. Y	69,422	90,903
Rochester, N. Y	62,386	89,363
Allegheny, Pa	53,180	78,681
Indianapolis, Ind	48,244	75,074
Richmond, Va	51,018	63,803
New Haven, Ct	50,840	62,882
Lowell, Mass	40,928	59,485
Worcester, Mass	41,105	58,295
Troy, N. Y	40,463	56,747
Kansas City, Mo	32,260	55,813
Cambridge, Mass	39,634	52,740
Syracuse, N. Y	43,051	51,791
Columbus, Ohio	31,274	51,665
Paterson, N. J	33,579	50,887
Toledo, Ohio	31,584	50,143
Charleston, S. C	48,956	49,999
Fall River, Mass	26,766	49,006
Minneapolis, Minn	13,066	46,887
Scranton, Pa	35,092	45,850
Nashville, Tenn	25,865	43,461
Reading, Pa	33,630	43,290
Hartford, Ct	37,180	42,553
Wilmington, Del	30,841	42,499
Camden, N. J	20,015	41,658
St. Paul, Minn	20,030	41,498
Lawrence, Mass	28,291	39,178
Dayton, Ohio	30,473	38,677
Lynn, Mass	28,233	38,284
Denver, Col	4,759	35,640
Oakland, Cal	10,500	34,556

AMERICAN INVENTORS.

Benjamin Franklin was born in Boston, Massachusetts, 1706; was the inventor of the lightning-rod; died in 1790.

Eli Whitney, born in Massachusetts 1765, and died 1825; invented the cotton-gin, 1793.

Robert Fulton, born in Pennsylvania 1765, died 1825; invented the steam-boat in 1793; in September, 1807, he ran the first steamboat, named the Clermont, from New York to Albany on the Hudson River; constructed the first steam war vessels in 1814.

J. Wood, born in New York 1774, died 1834; invented the cast-iron plow in 1814.

Thomas Blanchard, born in Massachusetts 1788, died 1864; invented the lathe and tack machines and constructed the stern-wheel boat.

Samuel Morse, born in Massachusetts 1791, died 1872; invented the magnetic telegraph; in 1835 the first wire was put up, one-half a mile long; to-day the miles of wire will number 250,000.

R. Winans, born in Maryland 1798; invented the pivoted, double-truck, long passenger coaches.

Charles Goodyear, born in Connecticut 1800; in 1839, by experimenting, he discovered a process for the manufacture of india-rubber, mixing sulphur and white lead with the native gum; from this invention we get our rubber goods, hose, and belting for machinery.

Cyrus McCormick, born in Virginia in 1809; invented the reaper in 1851.

Elias Howe, born in Massachusetts 1819, died

1867; is the original inventor of the sewing machine.

James Eads, born 1820; constructed and built the steel bridge over the Mississippi River at St. Louis in 1867; built many of the iron-proof steamers for the Government during the War of the Rebellion.

Cyrus W. Field, of New York, after spending millions of money and twelve years' labor, succeeded on the third trial in connecting the two hemispheres by a coil of wire laid in the bed of the ocean, the Atlantic cable, completed in 1866. Two years later the great Pacific Railroad was completed, a distance of two thousand miles, after six years' labor.

"MIXTURUM COMPLICATUM."

Frozen dew makes the frost.
Water boils at 212 deg. Fahrenheit.
Sound travels 1,120 feet per second.
Light moves 192,500 miles per second.
A cannon ball travels 1,560 feet per second.
A pigeon is said to fly 50 miles per hour.
Electricity moves 288,000 miles per second.
The earth moves in its orb 68,040 miles per hour.
A rifle-ball flies at the rate of 1,000 miles an hour.
Rain-water is considered the purest natural water.

The air balloon was invented by a Frenchman—Montgolfier—in 1782.

Washington's expenses during the Revolutionary War were $74,485.

The average weight for a horse to pull is estimated at 1,600 pounds.

The velocity of light was first determined by a Danish astronomer, Von Roemer.

The strength of an iron tube is greater than the same quantity made into a solid rod.

The influence of the heat of the sun is said to extend 50 feet into the earth.

The mean distance of the moon from the earth is 238,793 miles; its mean diameter is 2,160 miles.

The strength of a man is calculated to raise 10 pounds 10 feet high per second for 10 hours a day.

The light of the sun is estimated to be 300,000 times more intense than that of the moon.

The first railroad in the United States was in Quincy, Massachusetts, completed in 1827.

The greatest pyramid of Egypt is five hundred feet high; Herodotus says that it took 100,000 men twenty years to build it.

Slight winds travel one mile per hour; moderate winds, 7 miles; gales, 15 miles; high winds, 30 miles; storms, 50 miles; hurricanes, from 80 to 100 miles.

The strength of a horse is calculated to raise 33,000 pounds one foot high in a minute; by this steam engines are reckoned to be so many "horse-power."

It is said that Lord Rosse, of Ireland, constructed the largest telescope we have; it weighs four tons, is fifty feet long, and the diameter of the speculum is six feet.

A German chemist is said to have produced an artificial cold of 166 deg. (Fahrenheit) below zero, and yet failed to freeze pure alcohol. How is it to-day, my friends?

William Grier, of Pennsylvania, the original Garfield man, who "went it alone" in the Chicago Convention was tendered the place of Third Assistant Postmaster-General, April 20, 1881, but declined.

The first clock in England was made in A. D. 1288, and was considered so great a piece of ingenuity that a high dignitary was paid to take care of it.

In the Artesian wells of France, water rises from a depth

of 1,800 feet, to several feet above the surface. It is said that a well in Paris pours out 14,000,000 gallons per day.

Should two persons, one weighing 400 pounds, the other 100 pounds, fall from the same tower at the same moment, they would both strike the ground at the same instant. If you are any ways skeptical, try it.

General John A. Dix, of New York, at the breaking out of the Rebellion, issued an order to an official in Louisiana: "If any man attempts to haul down the American flag, shoot him on the spot."

Tides are caused by the attraction of the sun and moon; mostly by the latter. At each new moon the sun attracts in the same direction with the moon, when the tides are highest. The depths are also governed by surroundings. At St. Helena the tide rises to a depth of 3 feet, in the British Channel as high as 60 feet; and the highest being in the Bay of Fundy, known to reach 70 feet.

Each congressional district and territory in the United States is entitled to one cadet at the military school at West Point, formerly named by the representative, but later, left to a board of examiners for competition. Applicants must be between the ages of 17 and 22, sound, and not less than five feet in height; are required to serve the country four years after graduation.

"On Arch Street, Philadelphia, there stands to-day a house, No. 239. In this house was made the first flag of 'Stars and Stripes' by a Mrs. Ross; consisted of seven red and six white stripes alternating, and thirteen stars, representing the thirteen original states. It was intended that a stripe and star should be added whenever a new state was admitted into the Union, but in 1818 it was ordered by Congress that the flag should hereafter contain only thirteen stripes and that a star be added at the admission of each new state."

ENGLISH AUTHORS.

	Born.	Died.
Addison, Joseph	1672	1719
Akenside, Mark	1721	1770
Allison, Richard	—	—
Arnold, Edwin	1831	—
Arnold, Matthew	1822	—
Barbauld, Anna (nee Aikin)	1743	1825
Barham, Richard H.	1788	1845
Banes, William	1810	—
Barnfield, Richard	1574	1606
Barton, B	1784	1849
Baxter, Richard	1615	1691
Blake, William	1757	1827
Browning, Elizabeth Barrett	1805	1861
Browning, Robert	1812	—
Bulwer, Edward R.	1831	—
Bulwer, Edward G.	1805	1873
Byron, Lord George Gordon	1788	1824
Beaumont, Francis	1586	1616
Bloomfield, Robert	1766	1823
Cowper, William	1731	1800
Chatterton, Thomas	1752	1770
Clough, Arthur H.	1819	1861
Coleridge, Sam'l Taylor	1772	1834
Collins, William	1720	1756
Chaucer, Goeffrey	1328	1400
Crowe, Catherine (nee Stevens)	1800	—
Cook, Eliza *	1817	—
Canning, George	1770	1827
Caren, Thomas	1589	1639
Carlyle, Thomas	1795	1881
Dickens, Charles	1812	1870
Disraeli, Benjamin	1805	1881
Dryden, John	1631	1700
Edgeworth, Maria	1767	1849
Gray, Thomas	1716	1771
Gibbon, Edward	1737	1794
Hemans, Felicia (nee Browne)	1794	1835
Herbert, George	1593	1632
Hood, Thomas	1798	1845
Howitt, William	1795	—
Hunt, Leigh	1784	1859
Ingelow, Jean	1830	—
Jonson, Ben	1574	1637
Jones, Sir William	1746	1794
Keats, John	1796	1821
King, Henry	1591	1669
Kingsley, Charles	1819	1875
Lamb, Charles	1775	1834
Lovelace, Richard	1618	—
Macaulay, Lord Thomas B	1800	1859
Milton, John	1608	1674
Muloch, Dinah M.	1826	—
Mudford, William	1782	1845
Norton, Caroline (nee Sheridan)	1808	1877
Nash, Thomas	1558	1600
Pope, Alexander	1688	1744
Proctor, Adelaide A.	1825	1864
Proctor, Bryan Walter	1787	1874
Raleigh, Sir Walter	1552	1618
Reade, Charles	1814	—
Realf, Richard	1834	—
Richardson, Samuel	1689	1761
Ruskin, John	1819	—
Rogers, Samuel	1763	1855
Sedley, Sir Charles	1631	1701
Shakspeare, William	1564	1616
Sharpe, R. S	1759	1835
Shelley, Percy B	1792	1822
Shirly, James	1594	1666
Sidney, Sir Philip	1554	1586
Skelton, John	1485	1529
Smith, Charlotte	1749	1806
Smith, Horace	1779	1849
Smith, James	1776	1839
Smith, Sidney	1771	1845
Southey, Caroline (nee Bowles)	1787	1854
Southey, Robert	1774	1843
Spencer, William R.	1770	1834
Spencer, Edmund	1553	1599
Sterne, Laurence	1713	1768
Stevens, George A.	1784	—
Surrey, Lord	1516	1547
Tennyson, Alfred	1809	—
Thackeray, Wm. Makepeace, b. Calcutta	1811	1863
Tompson, James	1700	1748
Watts, Isaac	1674	1749
Wesley, Charles	1708	1788
Wordsworth, William	1770	1850

* Eliza Cook composed the poem, "The Old Arm-Chair." Caroline E. Norton composed the beautiful poem, "Bingen on the Rhine."

Benjamin Disraeli, the Earl of Beaconsfield, was born in London, 1805, and died April 20, 1881. The public career of this man was a most remarkable one; few men seldom attain high and honored positions in life, whose views, as changeable as the winds, are shaped to suit the exigencies of the times; this was Disraeli; and his success must be attributed to accident backed with a powerful genius. At the age of 22 he published his first novel, which embodied a display of wit and intellect that at once brought him into prominence. Three times he ran for a seat in Parliament and three times failed; the next time he succeeded. In a speech during a canvass he said:

"A statesman is the creature of his age, the child of circumstance, the creature of his times. A statesman is essentially a practical character, and when he is called upon to take office he is not to inquire what his opinions might or might not have been upon this or that subject; he is only to ascertain the needful and the beneficial, and the most feasible manner in which affairs are to be carried on. I laugh at the objection urged against a man that at a former period of his career he advocated a policy different to his present one. All I seek to ascertain is whether his present policy be just, necessary, expedient; whether at the present moment he is prepared to serve the country according to its present necessities."

If this little argument must blot out of memory the inconsistency of man's previous actions and put him on a new footing, and some "Yankee" held the document on sale under a patent, would not the Royal Bloods of America respond liberally, and would not the coffers of the "Yankee" weigh heavily with gold?

At the age of 32 he delivered his maiden speech in

the House of Commons; but through a noisy derision of the Whigs and embarrassment he failed, and upon taking his seat remarked: "I sit down now, but the time will come when you will hear me." About this time he told Lord Melbourne that he was going some day to be the Prime Minister of England. Ten years later he became the acknowledged leader of the great Conservative party of England, and by resignation of Gladstone in 1874, sure enough, became the Prime Minister of England, which position he held until his death. He was succeeded by Sir Stafford Northcote, or Lord Salisbury, May 9, 1881.

SCOTCH AUTHORS.

Name	Born	Died	Name	Born	Died
Aytoun, Sir Robert	1570	1638	Hogg, James	1770	1835
Beattie, James	1735	1803	Knox, William	1789	1825
Brown, John	1810	——	Leydon, John	1775	1811
Buchanan, Robert	1841	——	Mackenzie, Henry	1745	1831
Burns, Robert	1759	1796	Miller, Hugh	1802	1856
Campbell, Thomas	1777	1844	Montgomery, James	1771	1854
Gatt, John	1779	1839	Scott, Sir Walter	1771	1835
Hume, David	1711	1776	Sterling, John	1806	1844
Hervey, Thomas K	1779	1859	Wilson, John	1785	1854

IRISH AUTHORS.

Name	Born	Died	Name	Born	Died
Bannin, John	1798	1842	Griffin, Gerald	1803	1840
Carleton, William	1798	1869	Lover, Samuel	1797	1868
Davis, Thomas	1814	1845	McCarthy, Dennis F.	1810	——
Dufferin, Lady (nee Sheridan)	1807	1867	Moore, Thomas	1779	1852
			Wilde, Richard H	1789	1847
Goldsmith, Oliver	1728	1774	Wolfe, Charles	1719	1823

AUTHORS.

Name	Country	Born	Died
Angelo, Michel, Painter and Sculptor	Italy	1474	1563
Frei-Ligrath, Ferdinand, Lyric Poet	Germany	1810	1876
Girardin, St. Marc, Journalist	France	1801	1873
Goethe, Johann W., Poet	Germany	1749	1843
Gambetta, Leon, Statesman	France	1838	——

Heine, Heinrich, Poet and Wit	Germany	1799	1856
Luther, Martin, Leader of the Protestant Reformation	Germany	1483	1546
McMahon, Marshal, President of the French Republic, 1873–9	France	1808	——
Ronsard, Pierre de, a Reformer of French Poetry	France	1524	1585
Schiller, Johann C., Poet	Germany	1759	1805
Schelling, Friedrich, Philosopher	Germany	1775	1854
Theirs, Louis A., Historian, President French Republic, 1871–3	France	1797	1877

Francois P. Grevy, was born August 15, 1813; was elected in 1879 to the Presidency of the French Republic.

AMERICAN AUTHORS.

	Born.	Died.
Aldrich, James	1810	1856
Adams, John Quincy, Mass.	1767	1848
Alger, William R., Mass.	1823	——
Allston, Washington, S. C.	1779	1843
Andros, Richard S.	1800	1859
Barlow, Joel	1755	1812
Brown, Charles B.	1771	1810
Bryant, William Cullen, Mass.	1794	1878
Bancroft, George	1800	——
Baker, George H., Pa.	1824	——
Bolton, Sarah T., O.	——	——
Brainard, John G., Conn.	1796	1828
Brooks, Charles T. Mass.	1813	——
Brooks, Maria, Mass.	1795	1845
Brownell, Henry P.	1824	——
Bryant, John H., Mass	1807	——
Channing, William E.	1780	1842
Cooper, James Fennimore, N. J.	1789	1851
Carey, Alice, O.	1820	——
Carmichael, Sarah E.	——	——
Curtis, George W., R. I.	1824	——
Clark, George H.	——	——
Cooke, Philip P., Va.	1816	1850
Cutter, George W.	——	——
Dwight, Timothy	1752	1817
Dana, Richard Henry, Mass.	1787	1879
Drake, Joseph R., N. Y.	1795	1820
Davis, Rebecca H. (*nee*Harding), W. Va.	——	——
Dwight, John S.	1813	——
Eastman, Charles G., Me.	1816	1861
Emerson, Ralph Waldo, Mass.	1803	——
Edwards, Jonathan	1703	1758
Franklin, Benjamin	1706	1790
Freneau, Philip	1752	1832
Forrest, John W. De, Conn.	1826	——
Finch, Francis Miles, N. Y.	1827	——
Fenner, Cornelius G.	1822	1847
Fields, James T., N. H.	1820	——
Fosdick, William W., Ohio	1822	——
Glazier, William B., Me.	1827	——
Gould, Hannah F., Vt.	——	——
Hopkinson, Francis	1738	1791
Hawthorne, Nathaniel	1804	1864
Holmes, Oliver Wendell, Mass.	1809	——
Hale, Nathan, Conn.	1755	1776
Hale, Edward Everett, Mass.	1822	——

	Born.	Died.
Halleck, Fitz-Greene, Conn.	1790	1867
Harte, Bret, N. Y.	1839	—
Hoffman, Charles T., N. Y.	1806	—
Howells, William D., O.	1837	—
Hay, John, O.	1840	—
Hill, Thomas, N. J.	1818	—
Hillhouse, James A., Conn.	1789	1841
Holland, Josiah G., Mass.	1819	—
Howe, Julia Ward, N. Y.	1819	—
Hoyt, Ralph	—	—
Irving, Washington	1783	1859
Judson, Emily C. (nee Chabbuck), N. Y.	1817	1854
Kinney, Coates, N. Y.	1826	—
Key, Francis Scott, Md.	1779	1843
Longfellow, Henry Wadsworth, Me.	1807	—
Lowell, James Russell, Mass.	1819	—
Lowell, Robert T., Mass.	1816	—
Ludlow, Fitz-Hugh, N. Y.	1837	1870
Lowell, Maria W., Mass.	1821	1853
Motley, John Lothrop	1814	1877
Melville, Herman, N. Y.	1819	—
Mitchell, Donald G., Conn.	1822	—
Messenger, Robert H., Mass.	1807	—
Moore, Clement C., N. Y.	1779	1852
Morris, George P., Pa.	1802	1864
Muhlenberg, William A., N. Y.	—	—
Neal, John, Me.	1793	—
Osgood, Francis S., Mass.	1812	1850
Osgood, Kate Putnam, Me.	1843	—
O'Hara, Theodore, Ky.	1820	1867
O'Connor, William D., Mass.	1833	—
Paulding, James Kirke	1779	1860
Prescott, William H.	1796	1859
Poe, Edgar Allan, Md.	1809	1849
Parsons, Thomas W., Mass.	1819	—
Phelps, Elizabeth Stuart, Mass.	1844	—
Pierrepont, John, Conn.	1785	1866
Pinkney, Edward C., Md.	1802	1828
Palmer, John W., Md.	1825	—
Palmer, William Pitt, Mass.	1805	—
Payne, John Howard, N. Y.	1792	1852
Percy, Florence, Me.	1832	—
Priest, Nancy A.	1834	1870
Read, Thomas Buchanan, Pa.	1822	1872
Ridpath, John C., Ind.	1840	—
Randolph, Anson D., N. Y.	—	—
Raymond, Rossiter W., O.	1840	—
Saxe, John G., Vt.	1816	—
Sigourney, Lydia H.	1791	1865
Simmons, William G., S. C.	1806	—
Spencer, Caroline S., N. Y.	1850	—
Sprague, Charles, Mass.	1791	—
Stedman, Edmund C., Conn.	1833	—
Stoddard, Richard H., Mass.	1825	—
Spofford, Harriet P. (nee Prescott), Me.	1835	—
Story, William W., Mass.	1819	—
Stowe, Harriet Beecher, Conn.	1812	—
Tappan, William B., Mass.	1794	—
Taylor, Bayard, Pa.	1825	1878

	Born.	Died.
Taylor, Benjamin F. N. Y.	1822	—
Tilton, Theodore, N. Y.	1835	—
Trowbridge, John T., N. Y.	1827	—
Tuckerman, Henry T., Mass.	1813	—
Trumbull, John	1750	1831
Timrod, Henry, S. C.	1829	1867
Very, Jones, Mass.	1813	—
Wakefield, Nancy P. (nee Priest), N. H.	1837	1870
Walker, Elizabeth A., (nee Child), Vt.	—	—
Whittier, John Greenleaf, Mass.	1807	—
Willis, Nathaniel P., Me.	1807	1867
Winslow, Harriet	1824	—
Woodworth, Samuel, Mass.	1785	1842
Wilkinson, William C., Vt.	1833	—
Winthrop, Theodore, Conn.	1828	1861

Francis Bret Harte, author of poem "The Heathen Chinee."
Francis Scott Key,   "   "   "   "Star Spangled Banner."
Samuel Woodsworth,   "   "   "   "Old Oaken Bucket."
Thomas Buchanan Reed, "   "   "   "Sheridan's Ride."

THE HUMAN RACE.

The origin of the human race is said to have begun its existence in Asia, probably in the fertile valley of the Euphrates River. Many ages passed away during the dispersion of the people over the world, and to-day finds it inhabited in almost every section by people of various colors, customs, manners and habits of life. For distinction they have been classified into five varieties: The Caucasian, Ethiopian, Mongolian, American and Malay. The Caucasian stands at the head, including all the highly civilized nations of Europe and America—more conspicuous in arts, science and literature than any other race. To the Ethiopian class belongs the black man; to the Mongolian variety, the Chinese; with the American variety is classified the Indian or red man; while to the Malay belongs that class of people who inhabit the islands of the sea in the Indian and Pacific Oceans. Much has been written concerning the *national* difference of the races; many authors ascribe it to the climate, customs and surroundings, while there are some who consider the difference too great for a common origin, and hold that there were as many different pairs created in different localities by the Almighty as there are varieties. Among the writers of this class we find the eminent and no less distinguished naturalist, Professor Agassiz, of our time. He draws an analogy between the variety of animals and the races of men, claiming that there are zoological provinces for the different races of man, as there are for the different varieties of animals, and that the races, like animals, were separately created in their respective provinces; he discards the idea that animals were created in one part and then distributed; in support of his argument he remarks:

"How could the polar animals have migrated over the warmer tracts of land, which they would have to cross? for it is impossible now to keep them alive under such circumstances with the greatest precautions. And farther, some animals of the same species, sometimes presenting varieties and sometimes not, are found in different localities which are so cut off from all communication with each other that it is impossible that these animals could migrate from some one locality to all the rest. To assume that the geographical distribution of such animals, inhabiting zoological districts entirely disconnected with each other, is to be ascribed to physical causes, that these animals have been transported, and, especially, that the fishes which live in fresh water basins have been transported from place to place—to suppose that perches, pickerels, trouts and so many other species found in almost every brook and every river of the temperate zone, have been transported from one basin into another by freshets or by water birds—is to assume very inadequate and accidental causes for general phenomena. Not only then were different species of animals created originally in different localities, but it is also true, to a considerable extent, that animals of the same species, occupying different localities, were created in those localities." So he claims, in reference to the races, "there must have been, as in the case of animals, different original creations in the different zoological provinces," and holds that the history of Genesis refers to only one branch of the human family; but as the object of this work is to give facts, figures and *views*, without stopping to discuss the merits or demerits, we rest the question here.

THE ORIGIN OF SPOKEN AND WRITTEN LANGUAGE

It is claimed by quite a number of our best writers that spoken language is a divine gift from on high, while some maintain that man invented it for his own use. Among the advocates of the latter doctrine we find ancient writers like Horace, Cicero and Pliny; these authors claim that man reached his high, enlightened state by mere gradation; first, a rude, wandering, illiterate people, sensible of nothing but hunger and pain, communicating with each other after the manner of beasts. Lord Monboddo, a Scotch philosopher, claims that man at a very early time only used a few monosyllables in communicating with his fellowman, and considers man but a higher species of monkey. While this notion seems most absurd, as well as ridiculous, our observations of some specimens of humanity almost compel us to award a few grains of truth to his theory without further argument. From the Scriptures it is evident that Adam and Eve used a spoken language, hence the evidence seems conclusive that it is a divine institution. Probably it was an elementary language given to man for immediate use, which time, necessity and practice would greatly improve. In proof of this we find that such has been the case in the history of all nations; at first very imperfect and defective, but gradually improving as the races advanced to a higher state of civilization. Several centuries passed away, when man began to realize the necessity of a written language, that he might communicate with his friends at a distance. The first attempt was the introduction of the *Ideographic system*, which conveyed thoughts by the means of pictures; the hieroglyphics were a species of this system. An improvement followed in the *Verbal system*, in which characters were used to represent an object. The Chinese language, it is said, is written in this system. Next followed the *Syllabic system*, which employs a character to represent each syllable, but this, too, proved cumbersome, and was superseded by the *Alphabetic system*. This quite met the wants of man. By a combination of letters and syllables, words are formed that enable him to express abstract ideas, as well as material objects; to describe the various shades of thought and every emotion of the mind. It is claimed on good authority that the Phœnicians were the inventors of letters and first introduced them into Greece.

From the Greek alphabet is derived the Latin; from the Latin comes the English*; the English contains 26 letters; the Latin, 25; the French, 23; the Spanish, 27; the Russian, 41; the Hebrew, 22; the Italian, 20; the Persian, 32; the Sanscrit, 50; the Turkish, 33; the German, 26, etc.

* The letter *w* was added to the English, which did not appear in the Latin.

ORIGIN OF THE ENGLISH LANGUAGE.

In the land of England we find the origin of our language—now spoken by the most powerful people of all the human race. The earliest inhabitants of the British Islands of which we have any authentic account were a branch of the Celtic race—a rude, savage and uncivilized people. In the year 55 B. C., the Roman army, under Julius Cæsar, invaded the land of the Britons, conquered and subdued the Celts in the southern part of the Island (England) and introduced the language, customs and manners of Rome; a great portion of the Celts fled to Wales, Ireland and the mountains of Scotland. Those who remained accepted the situation and during four hundred years, under the subjugation of Rome, acquired some little degree of civilization and became a Latin-speaking people. At the beginning of the fifth century, when the barbarous Huns and Lombards began an invasion of Italy, which necessarily called the Roman troops home, the Roman-Britons were left defenseless. The Celts in the northern portions of the country, whither they had been driven by the Romans, sought this opportunity to regain their lost lands, and were fast accomplishing their designs when the Germans of Scandinavia,(Norway and Sweden) came to the rescue; they came, some authors say, by invitation of the Britons in the time of peril; others, that they only wished to take advantage of the absence of the Roman army, and under pretense of assisting the Romanized Britons, to drive out the Celts, then swallow the Britons and take the land for themselves. This latter version is probably the correct one, as these Germans, or Saxons, as they were known to the Britons, were considered the most fearless and renowned sea-navigators of their age, and *gloried* in fight, especially when plunder and booty were coupled to the *glory*. These German invasions continued until they had wrested all southern Britain from the Celts and driven them back into the northern regions —and this time to stay. Now it was that the Latin-Celtic language was supplanted by the Saxon tongue, which forms the true basis of our English language. The Angles and Frisians were contemporary settlers with these Saxons; hence the term Anglo-Saxon, and England is said to have derived its name from the Angles. The Saxons held to this language in all its purity for about six hundred years, when Britain was invaded by the Normans in 1066. The result of this conquest was the introduction of some five thousand new words and a modification of the Saxon vernacular. William, the Duke of Normandy, was now on the English throne; he set to work to wipe out the Anglo-Saxon and put in its stead the French language. All appointments were filled by the most learned Frenchmen, and the school-boys and girls required to translate their Latin lessons into French, ignoring, as far as possible, the Anglo-Saxon. Notwithstanding the endeavors of King William and his successors, the masses held to the Saxon tongue; so, for two hundred years of Norman supremacy, two distinct languages were spoken in the island, the Anglo-Saxon and the Norman-French. Dr. Smith, of London, in his work on the origin of the English language, says: "The most important change which converted the Anglo-Saxon into Old English, and which consists chiefly in the substitution of the vowel *e* for the different inflections, was not due in any considerable degree to the Norman conquest, though it was probably hastened by that event. It commenced even before the Norman conquest, and was owing to the same causes which led to similar changes in the kindred German dialects. The large introduction of French words into English dates from the time when the Normans began to speak the language of the conquered race. It is, however, an error to represent the English language as springing from a mixture of Anglo-Saxon and French; since a mixed language, in the strict sense of the term, may be pronounced an impossibility. The English still remained essentially a German tongue, though it received such large accessions of French words as materially to change its character. To fix with precision the date when this change took place is manifestly an impossible task. It was a gradual process, and must have advanced with more or less rapidity in different parts of the country. In remote and less frequented districts the mass of the population long preserved their pure Saxon speech."

In his work: "*An Introduction to the Literature of Europe,*" Henry Hallam says: "Nothing can be more difficult than to determine, except by an arbitrary line, the commencement of the English language; not so much, as in those of the Continent, because we are in want of materials, but rather from an opposite reason—the possibility of tracing a very gradual succession of verbal changes, that ended in a change of denomination. For when we compare the earliest English of the thirteenth century with the Anglo-Saxon of the twelfth, it seems hard to pronounce why it should pass for a separate language, rather than a modification of the former. We must conform, however, to usage, and say that Anglo-Saxon was converted into English: 1. By contracting or otherwise modifying the pronunciation and orthography of words; 2. By omitting many inflections, especially of the noun, and consequently making more use of articles and auxiliaries; and, 3. By the introduction of French derivatives."

The reader should bear in mind that the English, at the beginning of the fourteenth century, was quite different from the English of our present day, as the following verses from Chaucer will show:

> "The sleer of himself yet saugh I there,
> His herte-blood hath bathed al his here;
> The nayl y-dryve in the shode a-nyght;
> The colde deth, with mouth gapyng upright."

Standard authors have put the alterations in the language under the following epochs:

    I. *Anglo-Saxon*, from A. D. 450—1150.
    II. *Semi-Saxon*, from A. D. 1150—1250.
    III. *Old English*, from A. D. 1250—1350.
    IV. *Middle English*, from A. D. 1350—1550.
    V. *Modern English*, from A. D. 1550 to the present time.

The history of English literature really dates its origin with Chaucer.

There are now over forty thousand words in our language; twenty-three thousand are said to come from the Saxons, five thousand from the Norman French, and the remainder mostly from Latin; some from Greek, modern French and various other dialects.

An analysis of Scriptural passages and different works of standard authors shows the following result: Twenty-eight twenty-ninths of the Bible to be pure Saxon; seven-eighths of Milton; five-sixths of Shakspeare; four-fifths of Addison, and two-thirds of Pope and Hume.

ENGLISH AUTHORS.

Geoffrey Chaucer, the first ENGLISH POET, was born in England in 1328, and died in October, 1400 A. D. From his Canterbury Pilgrims we extract a specimen of his poetry:

> "A good man there was of religion,
> That was a poor PARSONE of a town;
> But rich he was in holy thought and work,
> He was also a learned man, a clerk,
> That Christ's gospel truly would preach.
> His parishens devoutly would he teach,
> Benigne he was and wondrous diligent,
> And in adversity full patient:
> And such he was yproved often times;
> Full loth were he to cursen for his tithes,
> But rather would he given, out of doubt,
> Unto his poor parishioners about,
> Of his offering, and eke of his substance;
> He could in little thing have suffisance.
> Wide was his parish, and houses far asunder,
> But he nor felt nor thought of rain or thunder,
> In sickness and in mischief to visit
> The farthest in his parish, much and oft,
> Upon his feet, and in his hand a staff.
> This noble ensample to his sheep he gave:
> That first he wrought, and afterward he taught,
> Out of the gospel he the words caught,
> And this figure he added yet thereto,
> That if gold rust, what should iron do?
> And if a priest be foul, on whom we trust,
> No wonder if a common man do rust;
> Well ought a priest ensample for to give,
> By his cleanness, how his sheep should live."

SIR JOHN DE MANDEVILLE, England, 1300—1371: He left England in 1322 and traveled in Persia, Egypt, Palestine and other Eastern countries. An account of his travels was published and dedicated to Edward

III. in 1356; this is said to be the first English prose book published.

EDMUND SPENCER was born in England in 1553; died in 1599; was educated at Cambridge University, and is the poet of poets between the days of Chaucer and the advent of Shakspeare. His Faery Queen is considered his best production.

SIR PHILIP SIDNEY, born in England in 1554, died in 1586. Dr. Smith says of him: "The jewel of the court, the darling of the people, and the liberal and judicious patron of arts and letters, his early and heroic death gave the crowning grace to a consummate character." He died of a wound received in battle while aiding the Protestants of the Netherlands in their struggle against the Spaniards.

SIR WALTER RALEIGH was born in England 1552 and died in 1618; brave, generous, courteous, and talented, he won the esteem of queen Elizabeth which he long maintained. In 1583, under commission of the queen, sent out his first expedition to colonize America. This proved a failure. Two or three others followed, and, like the first, were unsuccessful. After the death of the queen, James the sixth of Scotland ascended the throne in 1603, and became James the first of England, when England and Scotland were united under the name of Great Britain. Ireland included in 1801.

Sir Walter Raleigh, without positive proof, was accused of trying to seat Lady Arabella Stuart on the throne, whereof James had him imprisoned for twelve years under sentence of death, which was carried into execution in 1618; during his imprisonment he wrote a

History of the World, and the night before his execution composed the following verses:

> E'en such is time; which takes on trust
> Our youth, our joys, our all we have,
> And pays us but with earth and dust;
> Which in the dark and silent grave,
> When we have wandered all our ways,
> Shuts up the story of our days:
> But from this earth, this grave, this dust,
> My God shall raise me up I trust.

FRANCIS BACON was born in England 1561 and died in 1626. He assumed the task to learn his people how to *philosophize* rather than to teach Philosophy. Long before Aristotle, Logic was employed by a class of Logicians who styled themselves Sophists. About the year 400 B. C., in the days of Plato and Socrates, were the youths of Greece trained by these false teachers who put lies in the place of truth and by their assumed wisdom founded upon fallacies and sophisms, appeared to convert truth into error. Socrates came with his practical Philosophy—"Know thyself"— which he taught in the public streets and upon the highways until he had exposed and silenced the whole school of Sophists. His art of reasoning was founded upon "Common Sense," whether or not he brought "his Philosophy from Heaven to Earth" as Cicero declares. Aristotle following in the same vein of Plato and Socrates, with the addition of new rules and principles, framed and founded the inductive and deductive system of Logic that has become the universally accepted principle of reasoning. "He drew," says Dr. Coppee, "the true and somewhat nice distinction between Logic and Rhetoric, and established the fact (a fact not yet learned by many who call themselves logicians) that Logic is not concerned with the *truth* of propositions, but only with the reasoning upon such propositions as are given into its charge. If the premises be *true*, then Logic will give a *true* conclusion; but if the premises be *false*, Logic gives a *false* conclusion; but in this latter case the *Logic* is as good, and the argument as valid, as in the former." Since *deduction* proceeded from a general law to particular instances, Lord Bacon discarded the syllogism and applied *induction* as the new Logic of experimental Philosophy. "The starting point of Bacon's Philosophy," says Coppee, "was the assertion that the *universe is a great store-house of facts;* and that it is man's duty and interest, and it ought to be his pleasure, to explore, discover and understand these facts, not only in their isolated characters, but in their relations to each other and to the universe itself. His experiments and his use of the experiments of others, was to enable him to arrive at general laws of the universe. Now, corresponding with the world around us, that is, the world of Nature, there is a world within us,—the world of Thought. Let either be impaired or cease to exist, and in just such a proportion is the other impaired or does it cease to exist. To unite them we have sensation and perception, and the union is lost if sensation and perception fail. The happy union, then, of Thought and Nature would lead man to Truth, and to attain to truth is his highest aim. The various forms which truth assumes to inspire the faculties and entice the pursuits of men, are called sciences, and by an examination of multitudes of these phenomenal facts, the true definitions of the sciences might be made, their true relations determined, and a plan of classification formed for practical purposes. Such then, briefly, was the aim of the new experimental Philosophy proposed by Bacon in his *Instauratio Magna*. With it directly, Logic had but little to do; but that little led men of science into errors, which remain to the present day."

THE DRAMA.

The first dawn of our drama dates its origin just after the Norman Conquest, about 1100 A. D. That these theatrical plays were introduced by the French immediately after their conquest of Britain, there remains no doubt, since one of the earliest plays, that of *St. Catherine*, was performed in 1119 A. D. and acted in French. These plays consisted of *Mysteries* or *Miracles*, and were used by the clergy as a means of imparting religious instruction to the ignorant and uneducated laity of the twelfth century; they existed until the end of the fourteenth century, when they were superseded by a class of plays entitled a *Morality*. When civilization and learning began to spread among the masses, and it became known that somebody else knew something besides the preachers, the clergy who heretofore had monopolized the advantages of knowledge, were now fast losing their hold on the populace. While these dramatic representations were not divested entirely of a religious character, yet they partook more of a worldly spirit and portrayed a general expression of humanity; next came the farcical plays known as *Interludes*, which, at the beginning of the reformation, were acted upon the stage by the advocates of both religious sects—the Catholics and Lutherites—each ridiculing the doctrines of the other, or, by a common expression of the nineteenth century, "throwing dirt at each other."

About the middle of the sixteenth century dramatic plays were divided into two classes—*Comedy* and *Tragedy*.

These compositions were written and performed mostly by students of the universities on certain occasions or festivities of the kings. The first comedy in our language was a play entitled, "Ralph Royster Doyster," performed in London 1551; concerning this play Dr. Smith says: "The principal characters are a rich and pretty widow, her lover, and several suitors, the chief of whom is the foolish personage who gives the title to the play. This ridiculous pretender to gayety and love, a young heir just put into possession of his fortune, is surrounded by a number of intriguers and flatterers who pretend to be his friends, and who lead their dupe into all sorts of absurd and humiliating scrapes; and the piece ends with the return of the favored lover from a voyage which he had undertaken in a momentary pique. The manners represented are those of the middle class of the period, and the picture given of London citizen life in the middle of the sixteenth century is curious, animated and natural."

One of the earliest plays in tragedy was entitled, *Ferrex* and *Porrex*, written in blank verse, by Thomas Sackville, Lord Buckhurst and Thomas Norton, and performed in 1562 in the presence of queen Elizabeth.

WILLIAM SHAKSPEARE was born in the town of Stratford-on-Avon, Warwickshire, England, April 23, 1564; and died April 23d, 1616, at the age of 52 years.

Of his early education but little is known. As his father and mother were neither able to write, it is quite certain that he received no instruction from this source; but as there existed a "free grammar school" in Stratford, established for the education of the poor, it is quite likely that William attended this school and stored away such elementary instruction as it afforded.

At the age of twenty-two he resolved to leave his

native home, and enter a theatrical life in London; at the age of twenty-five he took stock in the company of the Globe—the first theatre in London; this company was composed of fifteen shareholders, and Shakspeare's name was the eleventh on the list. Shakspeare remained with the Globe company as an actor and shareholder for twenty-five years, and during that time wrote the greatest dramas of his life, and was recognized as the most powerful genius of all dramatic poets.

The Globe was a paying investment, as Shakspeare had purchased one hundred and seven acres of land with his earnings, besides his NEW PLACE in Stratford, which landed estate he had purchased in 1597, and fitted up anew, whither he retired from the turmoils of the stage in 1611, to pass the sunny side of life in ease and quiet repose.

The works of Shakspeare are classified into History and Fiction.

BEN JONSON is ranked next to Shakspeare as a writer of dramas, was educated at Cambridge, produced his first dramatic composition in 1596; was the author of many plays and numerous works; by request was buried in a vertical position in the churchyard of Westminster.

JOHN MILTON was born in London, December 9th, 1608, and died November 8th, 1674.

His literary career towers in grandeur above the poets of any age or nation; with unparalleled fertility of conception, in natural beauty and purity of sentiment, his sublimest poems *Paradise Lost* and *Paradise Regained* stand without a peer in all the literature of England.

JOHN DRYDEN, born in 1631, died in 1700; was the author of many dramas, poems and prose works.

JOHN BUNYAN, born in 1628, died in 1688; was a great writer of parables. Extremely pious, he identified himself with the Baptists; as this sect was opposed to the church of England at the time of the Restoration, the crown severely persecuted many of the leaders ; among them was Bunyan, whom they imprisoned for twelve years, during which time he composed his allegory, the *Pilgrim's Progress*. Upon the accession of James II. to the throne of England he was released, and became a leader in his adopted church.

ALEXANDER POPE, born in London 1688, died 1744. He began his literary career at the age of sixteen, wrote much and long, and had acquired considerable notoriety as a literary genius under the reign of queen Anne.

JOSEPH ADDISON, born 1672, died 1719. A new and different method of diffusing knowledge or literary efforts was brought about by Sir Richard Steele, in conjunction with whom Addison acted. They introduced periodicals which contained, in addition to their studied literature, the news items of the day. This new feature at once gave their periodicals a wonderful patronage. Steele was the founder of the tri-weekly *Tatler* in 1709, to which Addison was the constant contributor. In the year 1711 Addison and Steele issued the first copy of *The Spectator*. Now began the rise of journalism.

DANIEL DEFOE, born in 1661, died 1731. Like Bunyan, he was an ardent supporter of the dissenting sect, clinging fervently to the principles of constitutional liberty and advocating the principles of Protestantism so strenuously as to bring him not only into disfavor with the government, but his own life into jeopardy. He is considered the founder of the *English Novel;* is the author of *Robinson Crusoe*.

Sir WILLIAM BLACKSTONE, born in London 1723, died in 1780. His *Commentaries on the Laws of England* have ever been, and are to-day, considered the basis for all legal erudition.

ROBERT BURNS, Scotland's greatest poet, was born in 1759, and died in the year 1796. Among his best productions appears the "Cotter's Saturday Night." His works are satirical, descriptive, and lyric. Like Cowper and many other distinguished poets, he was given to dissipation, and died in the beginning of true greatness.

> "Gie fools their silks and knaves their wine,—
> A man's a man for a' that."

In the role of romantic writers appear none more magnificent than SIR WALTER SCOTT, who was born 1771 and died 1832; he produced some historical works, wrote the biographies of Dryden, Napoleon and others. The greatest work of this master genius was the production of the "Waverly Novels," which form an epoch in the history of modern literature.

"Richard Waverly had a facility in making long, dull speeches, consisting of truisms and common-places, hashed up with a technical jargon of office, which prevented the inanity (emptiness) of his orations from being discovered, and thus established the character of a profound politician."

"The seat of the muse is in the mist of the secret and solitary hill and her voice in the murmur of the mountain stream. He who wooes her must love the barren rock more than the fertile valley, and the solitude of the desert better than the festivity of the hall."— *Waverly.*

"Upon the eve of battle the western sky twinkled with stars, but a frost mist rising from the ocean covered the eastern horizon, and rolled in white wreaths along the plain where the adverse army lay couched upon their arms ; their camp-fires gleamed with an obscure and hazy lustre through the heavy fog which encircled them with a doubtful halo."—*Waverly.*

" Better a fool at a feast than a wise man at a fray."—*Ivanhoe.*

" I know not where the trick lies; but although I can enter an ordinary with as much audacity—rebuke the waiters and drawers as roundly, drink as deep a health, swear as round an oath, and fling my gold as freely about as any of the jingling spurs and white feathers that are around me,—yet hang me if I can ever catch the true grace of it, though I have practiced it an hundred times."—*Kenilworth.*

" Impudence is a commodity we must carry through the world with us;—I tell you my own stock of assurance was too small to trade upon; I was fain to take in a ton or two more of brass at every port where I touched in the voyage of life; and I started overboard what modesty and scruples I had remaining, in order to make room for the storage."—*Kenilworth.*

UNIVERSITIES AND COLLEGES IN THE UNITED STATES IN 1879

[From the American Almanac for 1881.]

STATES AND TERRITORIES.	No. of Colleges.	PREPARATORY DEPARTMENT.		COLLEGIATE DEPARTMENT.	
		No. of Instructors.	No. of Students.	No. of Instructors.	No. of Students.
Alabama	4	2	108	55	331
Arkansas	5	20	596	38	312
California	12	27	1,295	160	818
Colorado	2	2	70	13	39
Connecticut	3			130	924
Delaware	1	5	56	7	50
Georgia	7	8	278	47	602
Illinois	29	91	2,833	209	2,204
Indiana	15	30	1,624	118	1,039
Iowa	19	34	1,635	133	1,104
Kansas	8	9	699	62	373
Kentucky	14	12	614	116	1,161
Louisiana	7	13	509	36	277
Maine	3			36	440
Maryland	9	18	316	113	1,161
Massachusetts	7		50	145	1,982
Michigan	9	17	879	112	1,135
Minnesota	5	1	498	55	308
Mississippi	4	0	736	27	209
Missouri	15	46	1,305	166	1,559
Nebraska	4	2	504	21	113
Nevada	1	1	42	1	
New Hampshire	1			14	215
New Jersey	4	8	26	59	642
New York	29	115	3,102	470	3,531
North Carolina	8	7	356	63	906
Ohio	36	69	3,130	266	2,613
Oregon	8	17	701	24	252
Pennsylvania	28	57	203	308	2,040
Rhode Island	1			19	271
South Carolina	7	7	336	39	228
Tennessee	21	38	1,371	145	1,826
Texas	10	18	889	66	781
Vermont	2			16	120
Virginia	7	3	186	57	662
West Virginia	4	4	78	22	244
Wisconsin	8	20	881	93	773
Dist. of Columbia	4	11	211	35	145
Utah	1	3	325	3	
Washington	2		114	4	126
Total	364	735	28,456	3,506	31,616

COLLEGES FOUNDED.

HARVARD, the first and oldest college in the United States, was founded at Cambridge, Massachusetts, in the year 1638; Rev. Henry Dunster was the first President, 1640.

THE WILLIAM AND MARY COLLEGE was founded in Virginia, 1692.

YALE COLLEGE was founded in Connecticut in 1700, and established at New Haven in the year 1716.

NEW JERSEY COLLEGE was founded at Princeton, New Jersey, 1746.

THE UNIVERSITY OF PENNSYLVANIA was founded at Philadelphia, 1749.

COLUMBIA COLLEGE was founded in New York, 1754.

BROWN UNIVERSITY was founded in Rhode Island, 1764.

These were the first institutions of learning in America; as the Sciences, Art and Literature advanced, Colleges sprang into existence, until the nineteenth century finds nearly every state in the Union with its state University, surrounded by schools, seminaries and academies, not only in the large cities, but every town and hamlet, and even the rural districts, are blessed with the privilege of educating the youths of our land.

The first printing press in America was put up in Mexico in 1535; the first one in the United States was established at Cambridge, Massachusetts, 1639; its first publication was an almanac.

The first newspaper appeared in the United States at Boston, 1704, and was called the *News Letter*.

The first daily was issued at Philadelphia, 1784.

The first book published in the United States was in 1607, written by Captain John Smith; its title was "The True Relation of Virginia."

"TWINKLE, TWINKLE, LITTLE STAR."

Mica, mica, parva stella,
Miror, quænam sis tam bella!
Splendens eminus in illo
Alba velut gemma cœlo.
Quando fervens sol discessit
Nec calore prata pascit,
Mox ostendis lumen purum,
Micans, micans, per obscurum.
Tibi noctu qui vagatur,
Ob scintillulum gratatur;
Ni micares tu, non sciret,
Quas per vias errans iret.
Meum sæpe thalamam luce
Specularis curiosa,
Neque carpseris soporem,
Donec venit sol per auram.

From the Yale Courant.

FRANKLIN'S TOAST.

" Long after Washington's victories over the *French and English* had made his name familiar to all Europe, Dr. Franklin chanced to dine with the English and French ambassadors, when, as nearly as I can recollect the words, the following toasts were drunk: By the British ambassador: 'England—the *sun* whose bright beams enlighten and fructify the remotest corners of the earth.' The French ambassador, glowing with national pride, but too polite to dispute the previous toast, drank, 'France—the *moon*, whose mild, steady, and cheering rays are the delight of all nations: consoling them in darkness, and making their dreariness beautiful.' Doctor Franklin then arose, and with his usual dignified simplicity said, 'George Washington—the Joshua, *who commanded the sun and moon to stand still; and they obeyed him.*'"

SIGNS USED TO REPRESENT LETTERS IN THE MORSE TELEGRAPHIC SYSTEM.

ALPHABET.		NUMERALS.
a · —	n — ·	1 · — — — —
b — · · ·	o · · —	2 · · — — · ·
c · · · ·	p · · · · ·	3 · · · — ·
d — · ·	q · · — ·	4 · · · · —
e ·	r · · ·	5 — — —
f · — ·	s · · ·	6 · · · · · ·
g — — ·	t —	7 — — · ·
h · · · ·	u · · —	8 — · · · ·
i · ·	v · · · —	9 — · · — ·
j — · — ·	w · — —	0 — — — — —
k — · —	x · — · ·	
l —	y · · — ·	
m — —	z · · · ·	
	& · — · · ·	

Experienced operators take the message from sound. Morse's right of invention was at first disputed by other claimants; but an appeal to the courts sustained him, and Congress appropriated $30,000 to put the work under headway. The first public dispatch was sent over the wire in 1844, from Baltimore to Washington. The Democratic party was in convention at Baltimore May 27th; it was in this convention that the two-thirds rule was adopted, and James K. Polk was nominated. The news was transmitted to Washington on the 29th of May, 1844.

In the year 1876, A. G. Bell exhibits the invention of his Telephone at the Centennial; by means of this instrument, instead of transmitting intelligence by dots and dashes, the sounds of words are conveyed as they are spoken; while this invention will never supplant the Electro-Magnetic telegraph, it will, for moderate distances, become the adopted medium of communication for *every* branch of business; the number in use, July,

1881, in the United States will not fall short of one hundred thousand. In 1877 T. A. Edison invents the Phonograph.

And 1881 finds him solving the problem of making the blackness of night more brilliant than day by Electric-lighting.

UNITED STATES POSTAL REGULATIONS, AS REVISED UNDER ACT OF MARCH 3, 1879.

[From the American Almanac 1881.]

First Class Mail Matter.

LETTERS.—This class includes letters, postal cards, and anything sealed or otherwise closed against inspection or anything containing writing not allowed as an accompaniment to printed matter, under class three.

Postage, 3 cents each half ounce or fraction thereof.

On local or drop letters, at free delivery offices, 2 cents. At offices where no free delivery by carrier, 1 cent.

Prepayment by stamps invariably required.

Postal cards, 1 cent.

Registered letters, 10 cents in addition to the proper postage.

The Post-Office Department or its revenue is not by law liable for the loss of any registered mail matter.

Second Class.

REGULAR PUBLICATIONS.—This class includes all newspapers, periodicals, or matter exclusively in print and regularly issued at stated intervals as frequently as four times a year, from a known office of publication or news agency. Postage, 2 cents a pound or fraction thereof, prepaid by special stamps. Publications designed primarily for advertising or free circulation, or not having a legitimate list of subscribers, are excluded from the pound rate, and pay third class rates.

Third Class.

Mail matter of the third class includes books, transient newspapers and periodicals, circulars, and other matter wholly in print, legal and commercial papers filed out in writing, proof-sheets, corrected proof-sheets, and manuscript copy accompanying the same.

MS. unaccompanied by proof-sheets, letter rates.

Limit of weight, 4 pounds each package, except single books—weight not limited.

Postage, 1 cent for each 2 ounces, or fractional part thereof, invariably prepaid by stamps.

Class Fourth.

Embraces merchandise and all matter not included in the 1st, 2d, or 3d class, which is not liable to injure the mail matter. Limit of weight, 4 pounds.

Postage, 1 cent for each ounce or fractional part thereof, prepaid.

All packages of matter of the 3d or 4th class must be so wrapped or enveloped that their contents may be examined by postmasters without destroying the wrappers.

Matter of the second, third, or fourth class containing any writing, except as here specified, or except bills and receipts for periodicals, or printed commercial papers filled out in writing, as deeds, bills, etc., will be charged with letter postage; but the sender of any book may write names or addresses therein, or on the outside with the word "from" preceding the same, or may write briefly on any package the number and names of the articles inclosed.

Postal Money Orders.

An order may be issued for any amount, from one cent to fifty dollars, inclusive, but fractional parts of a cent cannot be included.

The fees for orders are:

On orders not exceeding $15 ....10 cents.
"   "   over $15 and not exceeding $30..15 "
"   "   over 30   "   "   40..20 "
"   "   over 40   "   "   50..25 "

When a larger sum than fifty dollars is required, additional orders must be obtained; but no more than three orders will be issued in one day from the same post-office to the same remitter in favor of the same payee.

Free Delivery.

The free delivery of mail matter at the residences of the people desiring it is required by law in every city of 50,000 or more population, and may be established at every place containing not less than 20,000 inhabitants. The present number of free delivery offices is ninety.

The franking privilege was abolished July 1, 1873, but the following mail matter may be sent free by legislative saving-clauses, viz.:

1. All public documents printed by order of Congress, the Congressional Record and speeches contained therein, franked by member of Congress or the Secretary of the Senate, or Clerk of the House.

2. Seeds transmitted by the Commissioner of Agriculture, or by any member of Congress, procured from that department.

3. All periodicals sent to subscribers within the county where printed.

4. Letters and packages relating exclusively to the business of the Government of the United States, mailed only by officers of the same, publications required to be mailed to the Librarian of Congress by the copyright law, and letters and parcels mailed by the Smithsonian Institution. All these must be covered by specially printed "penalty" envelopes or labels.

All communications to Government officers, and to or from members of Congress, are required to be prepaid by stamps.

THE DECLARATION OF INDEPENDENCE.

IN CONGRESS JULY 4, 1776.

WHEN, in the course of human events, it becomes necessary for one people to dissolve the political bands which have connected them with another, and to assume, among the powers of the earth, the separate and equal station to which the laws of nature and of nature's God entitle them, a decent respect to the opinions of mankind requires that they should declare the causes which impel them to the separation.

We hold these truths to be self-evident: that all men are created equal; that they are endowed, by their Creator, with certain unalienable rights; that among these are life, liberty, and the pursuit of happiness.

That to secure these rights, governments are instituted among men, deriving their just powers from the consent of the governed; that whenever any form of government becomes destructive to these ends, it is the right of the people to alter or to abolish it, and to institute a new government, laying its foundation on such principles, and organizing its powers in such form, as to them shall seem most likely to effect their safety and happiness. Prudence, indeed, will dictate that governments long established should not be changed for light and transient causes; and, accordingly, all experience hath shown, that mankind are more disposed to suffer while evils are sufferable, than to right themselves by abolishing the forms to which they are accustomed. But when a long train of abuses and usurpations, pursuing invariably the same object, evinces a design to reduce them under absolute despotism, it is their right, it is their duty, to throw off such government, and to provide new guards for their future security. Such has been the patient sufferance of these colonies; and such is now the necessity which constrains them to alter their former systems of government. The history of the present king of Great Britain is a history of repeated injuries and usurpations, all having in direct object the establishment of an absolute tyranny over these states. To prove this, let facts be submitted to a candid world.

He has refused his assent to laws the most wholesome and necessary for the public good.

He has forbidden his governors to pass laws of immediate and pressing importance, unless suspended in their operation, till his assent should be obtained; and when so suspended, he has utterly neglected to attend to them.

He has refused so pass other laws for the accommodation of large districts of people, unless those people would relinquish the right of representation in the legislature—a right inestimable to them, and formidable to tyrants only.

He has called together legislative bodies at places unusual, uncomfortable, and distant from the repository of their public records, for the sole purpose of fatiguing them into compliance with his measures.

He has dissolved representative houses repeatedly, for opposing, with manly firmness, his invasions on the rights of the people.

He has refused, for a long time after such dissolutions, to cause others to be elected; whereby the legislative powers, incapable of annihilation, have returned to the people at large, for their exercise, the state remaining, in the meantime, exposed to all the dangers of invasion from without, and convulsions within.

He has endeavored to prevent the population of these states; for that purpose obstructing the laws for naturalization of foreigners; refusing to pass others to encourage their migration hither, and raising the conditions of new appropriations of lands.

He has obstructed the administration of justice, by refusing his assent to laws for establishing judiciary powers.

He has made judges dependent on his will alone, for the tenure of their offices, and the amount and payment of their salaries.

He has erected a multitude of new offices, and sent hither swarms of officers, to harass our people, and eat out their substance.

He has kept among us in times of peace, standing armies, without the consent of our legislatures.

He has affected to render the military independent of, and superior to, the civil power.

He has combined with others to subject us to a jurisdiction foreign to our constitution, and unacknowledged by our laws: giving his assent to their acts of pretended legislation:

For quartering large bodies of armed troops among us:

For protecting them, by a mock trial, from punishment for any murders which they should commit on the inhabitants of these states:

For cutting off our trade with all parts of the world:

For imposing taxes on us without our consent:

For depriving us, in many cases, of the benefits of trial by jury:

For transporting us beyond seas to be tried for pretended offences:

For abolishing the free system of English laws in a neighboring province, establishing therein an arbitrary government, and enlarging its boundaries, so as to render it at once an example and fit instrument for introducing the same absolute rule into these colonies:

For taking away our charters, abolishing our most valuable laws, and altering, fundamentally, the forms of our governments:

For suspending our own legislatures, and declaring themselves invested with power to legislate for us in all cases whatsoever

He has abdicated government here, by declaring us out of his protection, and waging war against us.

He has plundered our seas, ravaged our coasts, burnt our towns, and destroyed the lives of our people.

He is at this time transporting large armies of foreign mercenaries to complete the works of death, desolation, and tyranny, already begun with circumstances of cruelty and perfidy scarcely paralleled in the most barbarous ages, and totally unworthy the head of a civilized nation.

He has constrained our fellow-citizens, taken captive on the high seas, to bear arms against their country, to become the executioners of their friends and brethren, or to fall themselves by their hands.

He has excited domestic insurrections among us, and has endeavored to bring on the inhabitants of our frontiers the merciless Indian savages, whose known rule of warfare is an undistinguished destruction of all ages, sexes, and conditions.

At every stage of these oppressions we have petitioned for redress in the most humble terms: our repeated petitions have been answered only by repeated injury. A prince, whose character is thus marked by every act which may define a tyrant, is unfit to be the ruler of a free people.

Nor have we been wanting in attentions to our British brethren. We have warned them, from time to time, of attempts by their legislature to extend an unwarrantable jurisdiction over us. We have reminded them of the circumstances of our emigration and settlement here. We have appealed to their native justice and magnanimity, and we have conjured them by the ties of our common kindred to disavow these usurpations, which would inevitably interrupt our connections and correspondence. They, too, have been deaf to the voice of justice and of consanguinity. We must, therefore, acquiesce in the necessity which denounces our separation, and hold them, as we hold the rest of mankind—enemies in war; in peace, friends.

We, therefore, the representatives of the UNITED STATES OF AMERICA, in general congress assembled, appealing to the Supreme Judge of the world for the rectitude of our intentions, do, in the name and by the authority of the good people of these colonies, solemnly publish and declare, That these United Colonies are, and of right ought to be, FREE and INDEPENDENT STATES; that they are absolved from all allegiance to the British crown, and that all political connection between them and the state of Great Britain is, and ought to be, totally dissolved; and that, as FREE and INDEPENDENT STATES, they have full power to levy war, conclude peace, contract alliances, establish commerce, and to do all other acts and things which INDEPENDENT STATES may of right do. And for the support of this Declaration, with a firm reliance on the protection of DIVINE PROVIDENCE, we mutually pledge to each other our lives, our fortunes, and our sacred honor.

JOHN HANCOCK.

New Hampshire.—Josiah Bartlett, William Whipple, Matthew Thornton.

Massachusetts Bay.—Samuel Adams, John Adams, Robert Treat Paine, Elbridge Gerry.

Rhode Island, etc.—Stephen Hopkins, William Ellery.

Connecticut.—Roger Sherman. Samuel Huntingdon, William Williams, Oliver Wolcott.

New York.—William Floyd, Philip Livingston, Francis Lewis, Lewis Morris.

New Jersey.—Richard Stockton, John Witherspoon, Francis Hopkinson, John Hart, Abraham Clark.

Pennsylvania.—Robert Morris, Benjamin Rush, Benjamin Franklin, John Morton, George Clymer, James Smith, George Taylor, James Wilson, George Ross.

Delaware.—Cæsar Rodney, George Read, Thos. M'Kean.

Maryland.—Samuel Chase, William Paca, Thomas Stone, Charles Carroll of Carrollton.

Virginia.—George Wythe, Richard Henry Lee, Thomas Jefferson, Benjamin Harrison, Thomas Nelson, Jr., Francis Lightfoot Lee, Carter Braxton.

North Carolina.—William Hooper, Joseph Hewes, John Penn.

South Carolina.—Edward Rutlege, Thomas Hayward, Jr., Thomas Lynch, Jr., Arthur Middleton.

Georgia.—Button Gwinnett, Lyman Hall, George Walton.

THE CONSTITUTION OF THE UNITED STATES.

We, the People of the United States, in order to form a more perfect union, establish justice, insure domestic tranquillity, provide for the common defense, promote the general welfare, and secure the blessings of liberty to ourselves and our posterity, do ordain and establish this Constitution for the United States of America.

ARTICLE I.—SECTION I.

1. All legislative powers herein granted, shall be vested in a congress of the United States, which shall consist of a senate and house of representatives.

SECTION II.

1. The house of representatives shall be composed of members chosen every second year by the people of the several states; and the electors in each state shall have the qualifications requisite for electors of the most numerous branch of the state legislature.

2. No person shall be a representative who shall not have attained to the age of twenty-five years, and been seven years a citizen of the United States, and who shall not, when elected, be an inhabitant of that State in which he shall be chosen.

3. Representatives and direct taxes shall be apportioned among the several states which may be included within this union, according to their respective numbers, which shall be determined by adding to the whole number of free persons including those bound to service for a term of years, and excluding Indians not taxed, three-

fifths of all other persons. The actual enumeration shall be made within three years after the first meeting of the congress of the United States, and within every subsequent term of ten years, in such manner as they shall by law direct. The number of representatives shall not exceed one for every thirty thousand, but each state shall have at least one representative; and until such enumeration shall be made, the state of New Hampshire shall be entitled to choose three; Massachusetts, eight; Rhode Island and Providence Plantations, one; Connnecticut, five; New York, six; New Jersey, four; Pennsylvania, eight; Delaware, one; Maryland, six; Virginia, ten; North Carolina, five; South Carolina, fine; and Georgia, three.

4. When vacancies happen in the representation from any state, the executive authority thereof shall issue writs of election to fill up such vacancies.

5. The house of representatives shall choose their speaker and other officers, and shall have the sole power of impeachment.

SECTION III.

1. The senate of the United States shall be composed of two senators from each state, chosen by the legislature thereof, for six years; and each senator shall have one vote.

2. Immediately after they shall be assembled in consequence of the first election, they shall be divided, as equally as may be, into three classes. The seats of the senators of the first class shall be vacated at the expiration of the second year; of the second class, at the expiration of the fourth year; and of the third class, at the expiration of the sixth year: so that one-third

may be chosen every second year; and if vacancies happen, by resignation or otherwise, during the recess of the legislature of any state, the executive thereof may make temporary appointments until the next meeting of the legislature, which shall then fill such vacancies.

3. No person shall be a senator who shall not have attained to the age of thirty years, and been nine years a citizen of the United States, and who shall not, when elected, be an inhabitant of that state for which he shall be chosen.

4. The vice-president of the United States shall be president of the senate, but shall have no vote unless they be equally divided.

5. The senate shall choose their other officers, and also a president *pro tempore*, in the absence of the vice-president, or when he shall exercise the office of president of the United States.

6. The senate shall have the sole power to try all impeachments. When sitting for that purpose, they shall be on oath or affirmation. When the president of the United States is tried, the chief justice shall preside; and no person shall be convicted without the concurrence of two-thirds of the members present.

7. Judgment, in case of impeachment, shall not extend further than to removal from office, and disqualification to hold and enjoy any office of honor, trust, or profit under the United States; but the party convicted shall nevertheless be liable and subject to indictment, trial, judgment, and punishment according to law.

SECTION IV.

1. The times, places, and manner of holding elections for senators and representatives, shall be prescribed in

each state by the legislature thereof; but the congress may, at any time, by law, make or alter such regulations, except as to the places of choosing senators.

2. The congress shall assemble at least once in every year, and such meeting shall be on the first Monday in December, unless they shall by law appoint a different day.

SECTION V.

1. Each house shall be the judge of the elections, returns, and qualifications of its own members; and a majority of each shall constitute a quorum to do business; but a smaller number may adjourn from day to day, and may be authorized to compel the attendance of absent members, in such manner and under such penalties as each house may provide.

2. Each house may determine the rules of its proceedings, punish its members for disorderly behavior, and, with the concurrence of two-thirds, expel a member.

3. Each house shall keep a journal of its proceedings, and from time to time publish the same, excepting such parts as may in their judgment require secrecy; and the yeas and nays of the members of either house, on any question, shall, at the desire of one-fifth of those present, be entered on the journal.

4. Neither house, during the session of congress, shall, without the consent of the other, adjourn for more than three days nor to any other place than that in which the two houses shall be sitting.

SECTION VI.

1. The senators and representatives shall receive a compensation for their services, to be ascertained by law, and paid out of the treasury of the United States.

They shall, in all cases, except treason, felony, and breach of the peace, be privileged from arrest during their attendance at the session of their respective houses, and in going to and returning from the same; and for any speech or debate in either house, they shall not be questioned in any other place.

2. No senator or representative shall, during the time for which he was elected, be appointed to any civil office under the authority of the United States which shall have been created, or the emoluments whereof shall have been increased, during such time; and no person holding any office under the United States shall be a member of either house during his continuance in office.

SECTION VII.

1. All bills for raising revenue shall originate in the house of representatives; but the senate may propose or concur with amendments, as on other bills.

2. Every bill which shall have passed the house of representatives and the senate, shall, before it become a law, be presented to the president of the United States: if he approve, he shall sign it; but if not, he shall return it, with his objections, to that house in which it shall have originated, who shall enter the objection at large on their journal, and proceed to reconsider it. If, after such reconsideration, two-thirds of that house shall agree to pass the bill, it shall be sent, together with the objections, to the other house, by which it shall likewise be reconsidered, and if approved by two-thirds of that house, it shall become a law. But in all such cases, the votes of both houses shall be determined by yeas and nays, and the names of the persons voting for and against the bill shall be entered on the journal of each house respectively.

If any bill shall not be returned by the president within ten days (Sundays excepted) after it shall have been presented to him, the same shall be a law in like manner as if he had signed it, unless the congress by their adjournment prevent its return, in which case it shall not be a law.

3. Every order, resolution, or vote, to which the concurrence of the senate and house of representatives may be necessary, except on a question of adjournment, shall be presented to the president of the United States; and before the same shall take effect, shall be approved by him, or being disapproved by him, shall be repassed by two-thirds of the senate and house of representatives, according to the rules and limitations prescribed in the case of a bill.

SECTION VIII.

The congress shall have power—

1. To lay and collect taxes, duties, imposts, and excises; to pay the debts and provide for the common defense and general welfare of the United States; but all duties, imposts, and excises, shall be uniform throughout the United States:

2. To borrow money on the credit of the United States:

3. To regulate commerce with foreign nations, and among the several states, and with the Indian tribes:

4. To establish a uniform rule of naturalization, and uniform laws on the subject of bankruptcies throughout the United States.

5. To coin money, regulate the value thereof, and of foreign coin, and fix the standard of weights and measures:

6. To provide for the punishment of counterfeiting the securities and current coin of the United States:

7. To establish post offices and post roads:

8. To promote the progress of science and useful arts, by securing for limited times to authors and inventors the exclusive right to their respective writings and discoveries:

9. To constitute tribunals inferior to the supreme court:

10. To define and punish piracies and felonies committed on the high seas, and offenses against the laws of nations:

11. To declare war, grant letters of marque and reprisal, and make rules concerning captures on land and water:

12. To raise and support armies, but no appropriation of money to that use shall be for a longer term than two years:

13. To provide and maintain a navy:

14. To make rules for the government and regulation of the land and naval forces:

15. To provide for calling forth the militia to execute the laws of the union, suppress insurrections, and repel invasions.

16. To provide for organizing, arming, and disciplining the militia, and for governing such part of them as may be employed in the service of the United States, reserving to the states, respectively, the appointment of the officers, and the authority of training the militia according to the discipline prescribed by congress.

17. To exercise exclusive legislation in all cases whatsoever, over such district, not exceeding ten miles square, as may, by cession of particular states, and the

acceptance of congress, become the seat of government of the United States, and to exercise like authority over all places purchased by the consent of the legislature of the state in which the same shall be, for the erection of forts, magazines, arsenals, dock-yards, and other needful buildings; and,

18. To make all laws which shall be necessary and proper for carrying into execution the foregoing powers, and all other powers vested by this constitution in the government of the United States, or in any department or officer thereof.

SECTION IX.

1. The migration or importation of such persons as any of the states now existing shall think proper to admit, shall not be prohibited by the congress prior to the year one thousand eight hundred and eight, but a tax or duty may be imposed on such importation, not exceeding ten dollars for each person.

2. The privilege of the writ of *habeas corpus* shall not be suspended, unless when, in cases of rebellion or invasion, the public safety may require it.

3. No bill of attainder, or *ex post facto* law, shall be passed.

4. No capitation or other direct tax shall be laid, unless in proportion to the census or enumeration hereinbefore directed to be taken.

5. No tax or duty shall be laid on articles exported from any state. No preference shall be given by any regulation of commerce or revenue to the ports of one state over those of another; nor shall vessels bound to or from one state be obliged to enter, clear, or pay duties in another.

6. No money shall be drawn from the treasury but in consequence of appropriations made by law; and a regular statement and account of the receipts and expenditures of all public money shall be published from time to time.

7. No title of nobility shall be granted by the United States, and no person holding any office of profit or trust under them shall, without the consent of the congress, accept of any present, emolument, office, or title of any kind whatever, from any king, prince or foreign state.

SECTION X.

1. No state shall enter into any treaty, alliance, or confederation; grant letters of marque and reprisal; coin money; emit bills of credit; make anything but gold and silver coin a tender in payment of debts; pass any bill of attainder, *ex post facto* law, or law impairing the obligation of contracts; or grant any title of nobility.

2. No state shall, without the consent of congress, lay any imposts or duties on imports or exports, except what may be absolutely necessary for executing its inspection laws; and the net produce of all duties and imposts, laid by any state on imports or exports, shall be for the use of the treasury of the United States, and all such laws shall be subject to the revision and control of the congress. No state shall, without the consent of congress, lay any duty of tunnage, keep troops or ships of war in time of peace, enter into any agreement or compact with another state, or with a foreign power, or engage in war, unless actually invaded, or in such imminent danger as will not admit of delay.

ARTICLE II.—SECTION I.

1. The executive power shall be vested in a president of the United States of America. He shall hold his office during the term of four years, and, together with the vice-president, chosen for the same term, be elected as follows:

2. Each state shall appoint, in such manner as the legislature thereof may direct, a number of electors, equal to the whole number of senators and representatives to which the state may be entitled in the congress; but no senator or representative, or person holding an office of trust or profit under the United States, shall be appointed an elector.

[3. The electors shall meet in their respective states, and vote by ballot for two persons, of whom one at least shall not be an inhabitant of the same state with themselves. And they shall make a list of all the persons voted for, and of the number of votes for each; which list they shall sign and certify, and transmit sealed to the seat of the government of the United States, directed to the president of the senate. The president of the senate shall, in the presence of the senate and house of representatives, open all the certificates, and the votes shall then be counted. The person having the greatest number of votes shall be president, if such number be a majority of the whole number of electors appointed; and if there be more than one who have such majority, and have an equal number of votes, then the house of representatives shall immediately choose, by ballot, one of them for president; and if no person have a majority, then, from the five highest on the list, the said house shall, in like manner, choose the president. But, in choosing the president, the votes shall be taken by states the representation from each state having one vote; a quorum for this purpose shall consist of a member or members from two-thirds of the states, and a majority of all the states shall be necessary to a choice. In every case, after the choice of the president, the person having the greatest number of votes of the electors, shall be the vice-president. But if there should remain two or more who have equal votes, the senate shall choose from them, by ballot, the vice-president.*]

4. The congress may determine the time of choosing the electors, and the day on which they shall give their votes; which day shall be the same throughout the United States.

5. No person, except a natural born citizen, or a citizen of the United States at the time of the adoption of this constitution, shall be eligible to the office of president; neither shall any person be eligible to that office, who shall not have attained to the age of thirty-five years, and been fourteen years a resident within the United States.

6. In case of the removal of the president from office, or of his death, resignation, or inability to discharge the powers and duties of the said office, the same shall devolve on the vice-president, and the congress may, by law, provide for the case of removal, death, resignation, or inability, both of the president and vice-president, declaring what officer shall then act as president, and such officer shall act accordingly, until the disability be removed, or a president shall be elected.

7. The president shall, at stated times, receive for his services a compensation, which shall neither be increased nor diminished during the period for which he shall have been elected, and he shall not receive within that

---
*Altered, see amend. art. 12, page 173.

period any other emolument from the United States, or any of them.

8. Before he enter on the execution of his office, he shall take the following oath or affirmation:

9. "I do solemnly swear (or affirm) that I will faithfully execute the office of president of the United States, and will, to the best of my ability, preserve, protect, and defend the constitution of the United States."

SECTION II.

1. The president shall be commander-in-chief of the army and navy of the United States, and of the militia of the several states, when called into the actual service of the United States; he may require the opinion, in writing, of the principal officer in each of the executive departments, upon any subject relating to the duties of their respective offices; and he shall have power to grant reprieves and pardons for all offences against the United States, except in cases of impeachment.

2. He shall have power, by and with the advice and consent of the senate, to make treaties, provided two-thirds of the senators present concur; and he shall nominate, and, by and with the advice and consent of the senate, shall appoint ambassadors, other public ministers, and consuls, judges of the supreme court, and all other officers of the United States, whose appointments are not herein otherwise provided for, and which shall be established by law. But the congress may, by law, vest the appointment of such inferior officers as they think proper in the president alone, in the courts of law, or in the heads of departments.

3. The president shall have power to fill up all vacancies that may happen during the recess of the senate, by

granting commissions which shall expire at the end of their next session.

SECTION III.

1. He shall, from time to time, give to the congress information of the state of the union, and recommend to their consideration such measures as he shall judge necessary and expedient; he may, on extraordinary occasions, convene both houses, or either of them, and, in case of disagreement between them, with respect to the time of adjournment, he may adjourn them to such time as he shall think proper; he shall receive ambassadors and other public ministers; he shall take care that the laws be faithfully executed; and shall commission all the officers of the United States.

SECTION IV.

1. The president, vice-president, and all civil officers of the United States, shall be removed from office on impeachment for, and conviction of, treason, bribery, or other high crimes and misdemeanors.

ARTICLE III.—SECTION I.

1. The judicial power of the United States shall be vested in one supreme court, and in such inferior courts as the congress may, from time to time, ordain and establish. The judges, both of the supreme and inferior courts, shall hold their offices during good behavior; and shall, at stated times, receive for their services a compensation which shall not be diminished during their continuance in office.

SECTION II.

1. The judicial power shall extend to all cases in law and equity, arising under this constitution, the laws of

the United States, and treaties made, or which shall be made, under their authority; to all cases affecting ambassadors, other public ministers and consuls; to all cases of admiralty and maritime jurisdiction; to controversies to which the United States shall be a party; to controversies between two or more states; between a state and citizens of another state; between citizens of another state; between citizens of different states; between citizens of the same state claiming lands under grants of different states; and between a state, or the citizens thereof, and foreign states, citizens or subjects.

2. In all cases affecting ambassadors, other public ministers and consuls, and those in which a state shall be a party, the supreme court shall have original jurisdiction. In all the other cases before mentioned, the supreme court shall have appellate jurisdiction, both as to law and fact, with such exceptions, and under such regulations, as the congress shall make.

3. The trial of all crimes, except in cases of impeachment, shall be by jury; and such trial shall be held in the state where the said crimes shall have been committed; but when not committed within any state, the trial shall be at such place or places as the congress may, by law, have directed.

SECTION III.

1. Treason against the United States shall consist only in levying war against them, or in adhering to their enemies, giving them aid and comfort. No person shall be convicted of treason unless on the testimony of two witnesses to the same overt act, or on confession in open court.

2. The congress shall have power to declare the pun-

ishment of treason, but no attainder of treason shall work corruption of blood or forfeiture, except during the life of the person attainted.

ARTICLE IV.—SECTION I.

1. Full faith and credit shall be given in each state to the public acts, records, and judicial proceedings of every other state. And the congress may, by general laws, prescribe the manner in which such acts, records, and proceedings shall be proved, and the effect thereof.

SECTION II.

1. The citizens of each state shall be entitled to all privileges and immunities of citizens in the several states.

2. A person charged in any state with treason, felony, or other crime, who shall flee from justice, and be found in another state, shall, on demand of the executive authority of the state from which he fled, be delivered up, to be removed to the state having jurisdiction of the crime.

3. No person held to service or labor in one state under the laws thereof, escaping into another, shall, in consequence of any law or regulation therein, be discharged from such service or labor, but shall be delivered up on claim of the party to whom such service or labor may be due.

SECTION III.

1. New states may be admitted by the congress into this union; but no new state shall be formed or erected within the jurisdiction of any other state; nor any state be formed by the junction of two or more states, or parts of states, without the consent of the legislature of the states concerned, as well as of the congress.

2. The congress shall have power to dispose of, and make all needful rules and regulations respecting the territory or other property belonging to the United States; and nothing in this constitution shall be so construed as to prejudice any claims of the United States, or of any particular state.

SECTION IV.

1. The United States shall guarantee to every state in this union a republican form of government, and shall protect each of them against invasion, and on application of the legislature, or of the executive, when the legislature can not be convened, against domestic violence.

ARTICLE V.

1. The congress, whenever two-thirds of both houses shall deem it necessary, shall propose amendments to this constitution, or, on the application of the legislatures of two-thirds of the several states, shall call a convention for proposing amendments, which, in either case, shall be valid, to all intents and purposes, as part of this constitution, when ratified by the legislatures of three-fourths of the several states, or by conventions in three-fourths thereof, as the one or the other mode of ratification may be proposed by the congress; provided, that no amendment which may be made prior to the year one thousand eight hundred and eight, shall in any manner affect the first and fourth clauses in the ninth section of the first article; and that no state, without its consent, shall be deprived of its equal suffrage in the senate.

ARTICLE VI.

1. All debts contracted and engagements entered into, before the adoption of this constitution, shall be as

valid against the United States under this constitution as under the confederation.

2. This constitution, and the laws of the United States which shall be made in pursuance thereof, and all treaties made, or which shall be made, under the authority of the United States, shall be the supreme law of the land; and the judges in every state shall be bound thereby, anything in the constitution or laws of any state to the contrary notwithstanding.

3. The senators and representatives before mentioned, and the members of the several state legislatures, and all executive and judicial officers, both of the United States and of the several states, shall be bound by oath or affirmation, to support this constitution: but no religious test shall ever be required as a qualification to any office or public trust under the United States.

ARTICLE VII.

1. The ratification of the conventions of nine states shall be sufficient for the establishment of this constitution between the states so ratifying the same.

Done in Convention, by the unanimous consent of the states present, the seventeenth day of September, in the year of our Lord one thousand seven hundred and eighty seven, and of the independence of the United States of America the twelfth. In witness whereof, we have hereunto subscribed our names.

GEORGE WASHINGTON,
President and deputy from Virginia.

New Hampshire.—John Langdon, Nicholas Gilman.
Massachusetts.—Nathaniel Gorham, Rufus King.
Connecticut.—William Samuel Johnson, Roger Sherman.

New York.—Alexander Hamilton.

New Jersey.—William Livingston, David Brearly, William Patterson, Jonathan Dayton.

Pennsylvania.—Benjamin Franklin, Thomas Mifflin, Robert Morris, George Clymer, Thomas Fitzsimons, Jared Ingersoll, James Wilson, Gouverneur Morris.

Delaware.—George Read, Gunning Bedford, jr., John Dickinson, Richard Bassett, Jacob Broom.

Maryland.—James McHenry, Daniel of St. Thomas Jenifer, Daniel Carroll.

Virginia.—John Blair, James Madison, jr.

North Carolina.—William Blount, Richard Dobbs Spaight, Hugh Williamson.

South Carolina.—John Rutledge, Charles Cotesworth Pinckney, Charles Pinckney, Pierce Butler.

Georgia.—William Few, Abraham Baldwin.

Attest: WILLIAM JACKSON, *Secretary.*

AMENDMENTS TO THE CONSTITUTION.

ARTICLE I.

1. Congress shall make no law respecting an establishment of religion, or prohibiting the free exercise thereof; or abridging the freedom of speech, or of the press; or the right of the people peaceably to assemble, and to petition the government for a redress of grievances.

ARTICLE II.

1. A well regulated militia being necessary to the security of a free state, the right of the people to keep and bear arms shall not be infringed.

ARTICLE III.

1. No soldier shall, in time of peace, be quartered in any house without the consent of the owner, nor in time of war but in a manner to be prescribed by law.

ARTICLE IV.

1. The right of the people to be secure in their persons, houses, papers, and effects, against unreasonable searches and seizures, shall not be violated, and no warrants shall issue but upon probable cause, supported by oath or affirmation, and particularly describing the place to be searched, and the persons or things to be seized.

ARTICLE V.

1. No person shall be held to answer for a capital or otherwise infamous crime, unless on a presentment or

indictment of a grand jury, except in cases arising in the land or naval forces, or in the militia when in actual service, in time of war or public danger; nor shall any person be subject, for the same offense, to be twice put in jeopardy of life or limb; nor shall be compelled in any criminal case to be a witness against himself; nor be deprived of life, liberty, or property, without due process of law; nor shall private property be taken for public use without just compensation.

ARTICLE VI.

1. In all criminal prosecutions, the accused shall enjoy the right to a speedy and public trial, by an impartial jury of the state and district wherein the crime shall have been committed, which district shall have been previously ascertained by law, and to be informed of the nature and cause of the accusation; to be confronted with the witnesses against him; to have compulsory process for obtaining witnesses in his favor; and to have the assistance of counsel for his defense.

ARTICLE VII.

1. In suits at common law, where the value in controversy shall exceed twenty dollars, the right of trial by jury shall be preserved; and no fact tried by a jury shall be otherwise re-examined in any court of the United States, than according to the rules at the common law.

ARTICLE VIII.

1. Excessive bail shall not be required, nor excessive fines imposed, nor cruel and unusual punishments inflicted.

ARTICLE IX.

1. The enumeration in the constitution of certain rights, shall not be construed to deny or disparage others retained by the people.

ARTICLE X.

1. The powers not delegated to the United States by the constitution, nor prohibited by it to the states, are reserved to the states respectively, or to the people.

ARTICLE XI.

1. The judicial power of the United States shall not be construed to extend to any suit in law or equity, commenced or prosecuted against one of the United States by citizens of another state, or by citizens or subjects of any foreign state.

ARTICLE XII.

1. The electors shall meet in their respective states, and vote by ballot for president and vice-president, one of whom, at least, shall not be an inhabitant of the same state with themselves; they shall name in their ballots the person voted for as president, and in distinct ballots the person voted for as vice-president; and they shall make distinct lists of all persons voted for as president, and of all persons voted for as vice-president, and of the number of votes for each, which lists they shall sign and certify, and transmit sealed to the seat of the government of the United States, directed to the president of the senate; the president of the senate shall, in the presence of the senate and house of representatives, open all the certificates, and the votes shall then be counted; the person having the greatest number of votes for president shall be the president, if such num-

ber be a majority of the whole number of electors appointed; and if no person have such majority, then from the persons having the highest numbers, not exceeding three, on the list of those voted for as president, the house of representatives shall choose immediately, by ballot, the president. But in choosing the president, the votes shall be taken by states, the representation from each state having one vote; a quorum for this purpose shall consist of a member or members from two-thirds of the states, and a majority of all the states shall be necessary to a choice. And if the house of representatives shall not choose a president, whenever the right of choice shall devolve upon them, before the fourth day of March next following, then the vice-president shall act as president, as in the case of the death or other constitutional disability of the president.

2. The person having the greatest number of votes as vice-president, shall be the vice-president, if such number be a majority of the whole number of electors appointed; and if no person have a majority, then from the two highest numbers on the list, the senate shall choose the vice-president; a quorum for the purpose shall consist of two-thirds of the whole number of senators, and a majority of the whole number shall be necessary to a choice.

3. But no person constitutionally ineligible to the office of president, shall be eligible to that of vice-president of the United States.

ARTICLE XIII.

Sec. 1. Neither slavery nor involuntary servitude, except as a punishment for crime, whereof the party shall have been duly convicted, shall exist within the United States, or any place subject to their jurisdiction.

Sec. 2. Congress shall have power to enforce this article by appropriate legislation.

ARTICLE XIV.

Sec. 1. All persons born or naturalized in the United States, and subject to the jurisdiction thereof, are citizens of the United States, and of the state wherein they reside. No state shall make or enforce any law which shall abridge the privileges or immunities of citizens of the United States; nor shall any state deprive any person of life, liberty, or property, without due process of law, nor deny to any person within its jurisdiction, the equal protection of the laws.

Sec. 2. Representatives shall be apportioned among the several states according to their respective numbers, counting the whole number of persons in each state, excluding Indians not taxed. But when the right to vote at any election for choice of electors for president and vice-president of the United States, representatives in congress, the executive and judicial officers of a state, or the members of the legislature thereof, is denied to any of the male inhabitants of such state being twenty-one years of age, and citizens of the United States, or in any way abridged, except for participation in rebellion or other crime, the basis of representation therein shall be reduced in the proportion which the number of such male citizens shall bear to the whole number of male citizens twenty-one years of age in such State.

Sec. 3. No person shall be a senator, or representative in congress, or elector of president and vice-president, or hold any office, civil or military, under the United States, or under any state, who, having previously taken an oath as a member of congress, or as an officer of the United States, or as a member of any state legislature, or as an executive or judicial officer of any state, to support the Constitution of the United States, shall have engaged in insurrection or rebellion against the same, or given aid and comfort to the enemies thereof; but congress may, by a vote of two-thirds of each house, remove such disability.

SEC. 4. The validity of the public debt of the United States authorized by law, including debts incurred for payments of pensions and bounties for services in suppressing insurrection or rebellion, shall not be questioned. But neither the United States nor any state shall assume or pay any debt or obligation incurred in aid of insurrection or rebellion against the United States, or any claim for the loss or emancipation of any slave; but all such debts, obligations, and claims, shall be held illegal and void.

SEC. 5. The congress shall have power to enforce, by appropriate legislation, the provisions of this article.

ARTICLE XV.

SEC. 1. The right of citizens of the United States to vote shall not be denied or abridged by the United States or by any state on account of race, color, or previous condition of servitude.

SEC. 2. The congress shall have power to enforce this article by appropriate legislation.

NOTE.—The first ten Amendments were proposed in 1789, and adopted in 1791; the eleventh was proposed in 1794, adopted in 1798; the twelfth, proposed in 1803 and adopted in 1804; the thirteenth proposed and adopted in 1865; the fourteenth proposed in 1866, adopted in 1868; the fifteenth proposed in 1869, adopted in 1870.

WASHINGTON'S FAREWELL ADDRESS.

SEPTEMBER 17, 1796.

Friends and fellow-citizens:

THE period for a new election of a citizen to administer the executive government of the United States being not far distant, and the time actually arrived when your thoughts must be employed in designating the person who is to be clothed with that important trust, it appears to me proper, especially as it may conduce to a more distinct expression of the public voice, that I should now apprise you of the resolution I have formed, to decline being considered among the number of those out of whom a choice is to be made.

I beg you, at the same time, to do me the justice to be assured, that this resolution has not been taken without a strict regard to all the considerations appertaining to the relation which binds a dutiful citizen to his country; and that, in withdrawing the tender of service, which silence in my situation might imply, I am influenced by no diminution of zeal for your future interest—no deficiency of respect for your past kindness; but am supported by a full conviction that the step is compatible with both.

The acceptance of, and continuance hitherto in, the office to which your suffrages have twice called me, have been a uniform sacrifice of inclination to the opinion of duty, and to a deference for what appeared to be your desire. I constantly hoped that it would have been much earlier in my power, consistently with motives which I was not at liberty to disregard, to return to that retirement from which I had been reluctantly drawn. The strength of my inclination to do this, previous to the last election, had even led to the preparation of an address to declare it to you; but mature reflection on the then perplexed and critical posture of affairs with foreign nations, and the unanimous advice of persons entitled to my confidence, impelled me to abandon the idea.

I rejoice that the state of your concerns, external as well as internal, no longer renders the pursuit of inclination incompatible with the sentiment of duty or propriety; and am persuaded, whatever partiality may be retained for my services, that in the present circumstances of our country, you will not disapprove of my determination to retire.

The impressions with which I first undertook the arduous trust were explained on the proper occasion. In the discharge of this trust, I will only say, that I have, with good intentions, contributed toward the organization and administration of the government the best exertions of which a very fallible judgment was capable. Not unconscious, in the outset, of the inferiority of my qualifications, experience in my own eyes, perhaps still more in the eyes of others, has strengthened the motives to diffidence of myself; and every day the increasing weight of years admonishes me more and more that the shade of retirement is as necessary for me as it will be welcome. Satisfied that, if any circumstances have given peculiar value to my services, they were temporary, I have the consolation to believe that, while choice and prudence invite me to quit the political scene, patriotism does not forbid it.

In looking forward to the moment which is to terminate the career of my political life, my feelings do not permit me to suspend the deep acknowledgment of that debt of gratitude which I owe to my beloved country, for the many honors it has conferred upon me; still more for the steadfast confidence with which it has supported me; and for the opportunities I have thence enjoyed, of manifesting my inviolable attachment, by services faithful and persevering, though in usefulness unequal to my zeal. If benefits have resulted to our country from these services, let it always be remembered in your praise, and as an instructive example in our annals, that under circumstances in which the passions, agitated in every direction, were liable to mislead, amid appearances somewhat dubious, vicissitudes of fortune often discouraging, in situations in which, not unfrequently, want of success has countenanced the spirit of criticism, the constancy of your support was the essential prop of the efforts, and a guarantee of the plans by which they were effected. Profoundly penetrated with this idea, I shall carry it with me to my grave, as a strong incitement to unceasing wishes that Heaven may continue to you the choicest tokens of its beneficence; that your union and brotherly affection may be perpetual; that the free constitution, which is the work of your hands, may be sacredly maintained; that its administration, in every department, may be stamped with wisdom and virtue; that, in fine, the happiness of the people of these states, under the auspices of liberty, may be made complete by so careful a preservation, and so prudent a use of this blessing, as will acquire to them the glory of recommending it to the applause, the affection and adoption of every nation which is yet a stranger to it.

Here, perhaps, I ought to stop; but a solicitude for your welfare, which can not end but with my life, and the apprehension of danger natural to that solitude, urge me, on an occasion like the present, to offer to your solemn contemplation, and to recommend to your frequent review, some sentiments, which are the result of much reflection, of no inconsiderable observation, and which appear to me all-important to the permanency of your felicity as a people. These will be offered to you with the more freedom, as you can only see in them the disinterested warnings of a parting friend who can possibly have no personal motive to bias his counsel; nor can I forget as an encouragement to it, your indulgent reception of my sentiments on a former and not dissimilar occasion.

Interwoven as is the love of liberty with every ligament of our hearts, no recommendation of mine is necessary to fortify or confirm the attachment.

The unity of government which constitutes you one people is also now dear to you. It is justly so; for it is a main pillar in the edifice of your real independence; the support of your tranquillity at home, your peace abroad; of your safety; of your prosperity; of that very liberty which you so highly prize. But as it is easy to foresee, that from different causes and from different quarters, much pains will be taken, many artifices employed, to weaken, in your minds, the conviction of this truth; as this is the point in your political fortress against which the batteries of internal and external enemies will be most constantly and actively (though often covertly and insidiously) directed, it is of infinite moment that you should properly estimate the immense value of your national union to your collective and individual happiness; that you should cherish a cordial, habitual and immovable attachment to it; accustoming yourselves to think and speak of it as a palladium of your political safety and prosperity; watching for its preservation with jealous anxiety; discountenancing whatever may suggest even a suspicion that it can, in any event, be abandoned; and indignantly frowning upon the first dawning of every attempt to alienate any portion of our country from the rest, or to enfeeble the sacred ties which now link together the various parts.

For this you have every inducement of sympathy and interest. Citizens by birth or choice of a common country, that country has a right to concentrate your affections. The name of *American*, which belongs to you in your national capacity, must always exalt the just pride of patriotism, more than any appellation derived from local discriminations. With slight shades of difference, you have the same religion, manners, habits, and political principles. You have, in a common cause, fought and triumphed together; the independence and liberty you possess are the work of joint councils and joint efforts, of common dangers, sufferings, and success.

But these considerations, however powerfully they address themselves to your sensibility, are generally outweighed by those which apply more immediately to your interest; here every portion of our country finds the most commanding motives for carefully guarding and preserving the union of the whole.

The north, in an unrestrained intercourse with the south, protected by the equal laws of a common government, finds in the productions of the latter great additional resources of maritime and commercial enterprise, and precious materials of manufacturing industry. The south, in the same intercourse, benefiting by the same agency of the north, sees its agriculture grow and its commerce expand. Turning partly into its own channels the seamen of the north, it finds its particular navigation invigorated; and while it contributes, in different ways, to nourish and increase the general mass of the national navigation, it looks forward to the protection of a maritime strength, to which itself is unequally adapted. The east, in like intercourse with the west, in the progressive improvement of interior communications by land and water, will more and more find a valuable vent for the commodities which it brings from abroad or manufactures at home. The west derives from the east supplies requisite to its growth and comfort; and what is perhaps of still greater consequence, it must, of necessity, owe the secure enjoyment of the indispensable outlets for its own productions, to the weight, influence and future maritime strength of the Atlantic side of the union, directed by an indissoluble community of interest as one nation. Any other tenure by which the west can hold this essential advantage, whether derived from its own separate strength, or from an apostate and unnatural connection with any foreign power, must be intrinsically precarious.

While, then, every part of our country thus feels an immediate and particular interest in union, all the parts combined can not fail to find, in the united mass of means and efforts, greater strength, greater resource, proportionably greater security from external danger, a less frequent interruption of their peace by foreign nations; and what is of inestimable value, they must derive from union an exemption from those broils and wars between themselves which so frequently afflict neighboring countries not tied together by the same government, which their own rivalships alone would be sufficient to produce, but which opposite foreign alliances, attachments, and intrigues would stimulate and embitter. Hence, likewise, they will avoid the necessity of those overgrown military establishments which, under any form of government, are inauspicious to liberty, and which are to be regarded as particularly hostile to republican liberty; in this sense it is that your union ought to be considered as a main prop of your liberty, and that the love of the one ought to endear to you the preservation of the other.

These considerations speak a persuasive language to every reflecting and virtuous mind, and exhibit the continuance of the UNION as a primary object of patriotic desire. Is there a doubt whether a common government can embrace so large a sphere? Let experience solve it. To listen to mere speculation in such a case were criminal. We are authorized to hope that a proper organization of the whole, with the auxiliary agency of governments for the respective subdivisions, will afford a happy issue of the experiment. It is well worth a fair and full experiment. With such powerful and obvious motives to union, affecting all parts of our country, while experience shall not have demonstrated its impracticability, there will always be reason to distrust the patriotism of those who, in any quarter, may endeavor to weaken its bands.

In contemplating the causes which may disturb our union, it occurs as matter of serious concern, that any ground should have been furnished for characterizing parties by geographical discriminations—northern and southern, atlantic and western—whence designing men may endeavor to excite a belief that there is a real difference of local interests and views. One of the expedients of party to acquire influence within particular districts, is to misrepresent the opinions and aims of other districts. You can not shield yourselves too much against the jealousies and heart-burnings which spring from these misrepresentations; they tend to render alien to each other those who ought to be bound together by fraternal affection. The inhabitants of our western country have lately had a useful lesson on this head; they have seen, in the negotiation by the executive, and in the unanimous ratification by the senate, of the treaty with Spain, and in the universal satisfaction at that event throughout the United States, a decisive proof how unfounded were the suspicions propagated among them of a policy in the general government and in the Atlantic states unfriendly in regard to the Mississippi; they have been witnesses to the formation of two treaties—that with Great Britain and that with Spain—which secure to them everything they could desire, in respect to our foreign relations, toward confirming their prosperity. Will it not be their wisdom to rely, for the preservation of these advantages, on the UNION by which they were procured? Will they not henceforth be deaf to those advisers, if such there are, who would sever them from their brethren and connect them with aliens?

To the efficacy and permanency of your union, a government for the whole is indispensable. No alliances, however strict, between the parts, can be an adequate substitute; they must inevitably experience the infractions and interruptions which alliances, in all times, have experienced. Sensible of this momentous truth, you have improved upon your first essay by the adoption of a constitution of government better calculated than your former for an intimate union, and for the efficacious management of your common concerns. This government, the offspring of your own choice, uninfluenced and unawed, adopted upon full investigation and mature deliberation, completely free in its principles, in the distribution of its powers, uniting security with energy, and containing within itself provision for its own amendment, has a just claim to your confidence and your support. Respect for its authority, compliance with its laws, acquiescence in its measures, are duties enjoined by the fundamental maxims of true liberty. The basis of our political system is the right of the people to make and to alter their constitutions of government; but the constitution which, at any time, exists, until changed by an explicit and authentic act of the whole people, is sacredly obligatory upon all. The very idea of the power and the right of the people to establish government, presupposes the duty of every individual to obey the established government.

All obstructions to the execution of the laws, all combinations and associations, under whatever plausible character, with the real design to direct, control, counteract, or awe the regular deliberation and action of the constituted authorities, are destructive of this fundamental principle, and of fatal tendency. They serve to organize faction, to give it an artificial and extraordinary force, to put in the place of the delegated will of the nation, the will of party, often a small but artful and enterprising minority of the community, and according to the alternate triumphs of different parties, to make the public administration the mirror of the ill-concerted and incongruous projects of faction, rather than the organ of consistent and wholesome plans, digested by common councils, and modified by mutual interests.

However combinations or associations of the above description may now and then answer popular ends, they are likely, in the course of time and things, to become potent engines by which cunning, ambitious and unprincipled men will be enabled to subvert the power of the people, and to usurp for themselves the reins of government, destroying afterward the very engines which have lifted them to unjust dominion.

Towards the preservation of your government and the permanency of your present happy state, it is requisite, not only that you steadily denounce irregular opposition to its acknowledged authority, but also that you resist with care the spirit of innovation upon its principles, however specious the pretexts. One method of assault may be, to effect, in the forms of the constitution, alterations which will impair the energy of the system, and thus to undermine what cannot be directly overthrown. In all the changes to which you may be invited, remember that time and habit are at least as necessary to fix the true character of governments as of other human institutions; that experience is the surest standard by which to test the real tendency of the existing constitutions of a country; that facility in changes, upon the credit of mere hypothesis and opinion, exposes to perpetual change, from the endless variety of hypothesis and opinion; and remember, especially, that, from the efficient management of your common interests, in a country so extensive as ours, a government of as much vigor as is consistent with the perfect security of liberty is indispensable. Liberty itself will find in such a government, with powers properly distributed and adjusted, its surest guardian. It is, indeed, little else than a name where the government is too feeble to withstand the enterprises of faction, to confine each member of society within the limits prescribed by the laws, and to maintain all in the secure and tranquil enjoyment of the rights of person and property.

I have already intimated to you the danger of parties in the state, with particular reference to the founding of them on geographical discriminations. Let me now take a more comprehensive view, and warn you, in the most solemn manner, against the baneful effects of the spirit of party generally.

This spirit, unfortunately, is inseparable from our nature, having its root in the strongest passions of the human mind. It exists, under different shapes, in all governments, more or less stifled, controlled, or repressed; but in those of the popular form it is seen in its greatest rankness, and is truly their worst enemy.

The alternate domination of one faction over another, sharpened by the spirit of revenge natural to party dissension, which, in different ages and countries, has perpetrated the most horrid enormities, is itself a frightful despotism. But this leads at length to a more formal and permanent despotism. The disorders and miseries which result, gradually incline the minds of men to seek security and repose in the absolute power of an individual, and, sooner or later, the chief of some prevailing faction, more able or more fortunate than his competitors, turns this disposition to the purposes of his own elevation on the ruins of the public liberty.

Without looking forward to an extremity of this kind

(which, nevertheless, ought not to be entirely out of sight, the common and continual mischiefs of the spirit of party are sufficient to make it the interest and duty of a wise people to discourage and restrain it.

It serves always to distract the public councils, and enfeeble the public administration. It agitates the community with ill-founded jealousies and false alarms; kindles the animosity of one part against another, foments occasional riot and insurrection. It opens the door to foreign influence and corruption, which finds a facilitated access to the government itself, through the channels of party passion. Thus the policy and will of one country are subjected to the policy and will of another.

There is an opinion that parties in free countries are useful checks upon the administration of the government, and serve to keep alive the spirit of liberty. This, within certain limits, is probably true; and in governments of a monarchical cast, patriotism may look with indulgence, if not with favor, upon the spirit of party. But in those of popular character, in governments purely elective, it is a spirit not to be encouraged. From the natural tendency, it is certain there will always be enough of that spirit for every salutary purpose. And there being constant danger of excess, the effort ought to be, by force of public opinion, to mitigate and assuage it. A fire not to be quenched, it demands a uniform vigilance to prevent its bursting into a flame, lest, instead of warming, it should consume.

It is important, likewise, that the habits of thinking in a free country should inspire caution in those intrusted with its administration, to confine themselves within their respective constitutional spheres, avoiding, in the exercise of the powers of one department, to encroach upon another. The spirit of encroachment tends to consolidate the powers of all the departments in one, and thus to create, whatever the form of government, a real despotism. A just estimate of that love of power and proneness to abuse it, which predominate in the human heart, is sufficient to satisfy us of the truth of this position. The necessity of reciprocal checks in the exercise of political power, by dividing and distributing it into different depositories, and constituting each the guardian of the public weal against invasions of the other, has been evinced by experiments, ancient and modern; some of them in our country, and under our own eyes. To preserve them must be as necessary as to institute them. If, in the opinion of the people, the distribution or modification of the constitutional powers be in any particular wrong, let it be corrected by an amendment in the way in which the constitution designates. But let there be no change by usurpation; for though this in one instance may be the instrument of good, it is the customary weapon by which free governments are destroyed. The precedent must always greatly overbalance in permanent evil any partial or transient benefit which the use can at any time yield.

Of all the dispositions and habits which lead to political prosperity, religion and morality are indispensable supports. In vain would that man claim the tribute of patriotism who should labor to subvert these great pillars of human happiness—these firmest props of the duties of men and citizens. The mere politician, equally with the pious man, ought to respect and to cherish them. A volume could not trace all their connection with private and public felicity. Let it be simply asked, where is the security for property, for reputation, for life, if the sense of religious obligation *desert* the oaths which are the instruments of investigation in courts of justice? And let us with caution indulge the supposition that morality can be maintained without religion. Whatever may be conceded to the influence of refined education on minds of peculiar structure, reason and experience both forbid us to expect that national morality can prevail in exclusion of religious principles.

It is substantially true that virtue or morality is a necessary spring of popular government. The rule indeed extends with more or less force to every species of free government. Who that is a sincere friend to it can look with indifference upon attempts to shake the foundation of the fabric?

Promote, then, as an object of primary importance, institutions for the general diffusion of knowledge. In proportion as the structure of a government gives force to public opinion, it is essential that public opinion should be enlightened.

As a very important source of strength and security, cherish public credit. One method of preserving it is to use it as sparingly as possible, avoiding occasions of expense by cultivating peace, but remembering, also, that timely disbursements to prepare for danger frequently prevent much greater disbursements to repel it; avoiding, likewise, the accumulation of debt, not only by shunning occasions of expense, but by vigorous exertions in time of peace to discharge the debts which unavoidable wars have occasioned, not ungenerously throwing upon posterity the burden which we ourselves ought to bear. The execution of these maxims belongs to your representatives, but it is necessary that public opinion should co-operate. To facilitate to them the performance of their duty, it is essential you should practically bear in mind that, toward the payment of debts, there must be revenue; that to have revenue there must be taxes; that no taxes can be devised which are not more or less inconvenient and unpleasant; that the intrinsic embarrassment inseparable from the selection of the proper objects (which is always a choice of difficulties), ought to be a decisive motive for a candid construction of the conduct of the government in making it, and for a spirit of acquiescence in the measures for obtaining revenue, which the public exigencies may at any time dictate.

Observe good faith and justice toward all nations; cultivate peace and harmony with all. Religion and morality enjoin this conduct; and can it be that good policy does not equally enjoin it? It will be worthy of a free, enlightened, and, at no distant period, a great nation, to give to mankind the magnanimous and too novel example of a people always guided by an exalted justice and benevolence. Who can doubt that in the course of time and things the fruits of such a plan would richly repay any temporary advantages that might be lost by a steady adherence to it? Can it be that Providence has connected the permanent felicity of a nation with its virtue? The experiment, at least, is recommended by every sentiment which ennobles human nature. Alas! it is rendered impossible by its vices.

In the execution of such a plan, nothing is more essential than that permanent, inveterate antipathies against particular nations and passionate attachments for others should be excluded; and that, in the place of them, just and amicable feelings toward all should be cultivated. The nation which indulges toward another an habitual hatred or an habitual fondness is in some degree a slave. It is a slave to its animosity or to its affection, either of which is sufficient to lead it astray from its duty and its interest. Antipathy in one nation against another disposes each more readily to offer insult and injury, to lay hold of slight causes of umbrage, and to be haughty and intractable when accidental or trifling occasions of dispute occur.

Hence, frequent collisions and obstinate, envenomed and bloody contests. The nation, prompted by ill-will and resentment, sometimes impels to war the government contrary to the best calculations of policy. The government sometimes participates in the national propensity, and adopts, through passion, what reason would reject. At other times it makes the animosity of the nation subservient to the projects of hostility, instigated by pride, ambition, and other sinister and pernicious motives. The peace often, sometimes perhaps the liberty, of nations has been the victim.

So, likewise, a passionate attachment of one nation for another produces a variety of evils. Sympathy for the favorite nation, facilitating the illusion of an imaginary common interest in cases where no real common interest exists, and infusing into one the enmities of the other, betrays the former into a participation in the quarrels and the wars of the latter, without adequate inducements or justification. It leads, also, to concessions to the favorite nation of privileges denied to others, which are apt doubly to injure the nation making the concessions by unnecessarily parting with what ought to have been retained, and by exciting jealousy, ill-will, and a disposition to retaliate in the parties from whom equal privileges are withheld; and it gives to ambitious, corrupt or deluded citizens, who devote themselves to the favorite nation, facility to betray or sacrifice the interests of their own country without odium, sometimes even with popularity, gilding with the appearances of a virtuous sense of obligation to a commendable deference for public opinion, or a laudable zeal for public good, the base or foolish compliances of ambition, corruption or infatuation.

As avenues to foreign influences in innumerable ways, such attachments are particularly alarming to the truly enlightened and independent patriot. How many opportunities do they afford to tamper with domestic factions, to practice the arts of seduction, to mislead public opinion, to influence or awe the public councils! Such an attachment of a small or weak nation toward a great and powerful one dooms the former to be the satellite of the latter. Against the insidious wiles of foreign influence, I conjure you to believe me, fellow-citizens, the jealousy of a free people ought to be constantly awake, since history and experience prove that foreign influence is one of the most baneful foes of republican government. But that jealousy, to be useful, must be impartial, else it becomes the instrument of the very influence to be avoided, instead of a defense against it. Excessive partiality for one foreign nation and excessive dislike for another cause those whom they actuate to see danger only on one side, and serve to veil and even second the arts of influence on the other. Real patriots, who may resist the intrigues of the favorite, are liable to become suspected and odious, while its tools and dupes usurp the applause and confidence of the people to surrender their interests.

The great rule of conduct for us in regard to foreign nations is, in extending our commercial relations, to have with them as little political connection as possible. So far as we have already formed engagements, let them be fulfilled with perfect good faith. Here let us stop.

Europe has a set of primary interests which to us have none or a very remote relation. Hence, she must be engaged in frequent controversies, the causes of which are essentially foreign to our concerns. Hence, therefore, it must be unwise in us to implicate ourselves by artificial ties in the ordinary vicissitudes of her politics, or the ordinary combinations and collisions of her friendships or enmities.

Our detached and distant situation invites and enables us to pursue a different course. If we remain one people, under an efficient government, the period is not far off when we may defy material injury from external annoyance; when we may take such an attitude as will cause the neutrality we may at any time resolve upon to be scrupulously respected; when belligerent nations, under the impossibility of making acquisitions upon us, will not lightly hazard the giving us provocation; when we may choose peace or war, as our interests, guided by justice, shall counsel.

Why forego the advantages of so peculiar a situation? Why quit our own to stand on foreign ground? Why, by interweaving our destiny with that of any part of Europe, entangle our peace and prosperity in the toils of European ambition, rivalship, interest, humor or caprice?

It is our true policy to steer clear of permanent alliances with any portion of the foreign world, so far, I mean, as we are now at liberty to do it; for let me not be understood as capable of patronizing infidelity to existing engagements. I hold the maxim no less applicable to public than to private affairs, that honesty is always the best policy. I repeat, therefore, let those engagements be observed in their genuine sense. But, in my opinion, it is unnecessary and would be unwise to extend them.

Taking care always to keep ourselves by suitable establishments on a respectable defensive posture, we may safely trust to temporary alliances for extraordinary emergencies.

Harmony and liberal intercourse with all nations are recommended by policy, humanity and interest. But even our commercial policy should hold an equal and impartial hand; neither seeking nor granting exclusive favors or preferences; consulting the natural course of things; diffusing and diversifying by gentle means the stream of commerce, but forcing nothing; establishing, with powers so disposed (in order to give trade a stable course, to define the rights of our merchants, to enable the government to support them), conventional rules of intercourse, the best that present circumstances and natural opinion will permit, but temporary, and liable to be, from time to time, abandoned or varied as experience or circumstances shall dictate; constantly keeping in view that it is folly in one nation to look for disinterested favors from another; that it must pay with a portion of its independence for whatever it may accept under that character; that by such acceptance it may place itself in the condition of having given equivalents for nominal favors, and yet of being reproached with ingratitude for not having given more. There can be no greater error than to expect or calculate upon real favors from nation to nation. It is an illusion which experience must cure, which a just pride ought to discard.

In offering to you, my countrymen, these counsels of an old, affectionate friend, I dare not hope that they will make the strong and lasting impression I could wish; that they will control the usual current of the passions, or prevent our nation from running the course which has hitherto marked the destiny of nations. But if I may even flatter myself that they may be productive of some partial benefit, some occasional good; that they may now and then recur to moderate the fury of party spirit, to warn against the mischiefs of foreign intrigue, to guard against the impostures of pretended patriotism; this hope will be a full recompense for the solicitude for your welfare by which they have been dictated.

How far, in the discharge of my official duties, I have been guided by the principles which have been delineated, the public records and the other evidences of my conduct must witness to you and to the world. To myself, the assurance of my own conscience is that I have at least believed myself to be guided by them.

In relation to the still subsisting war in Europe, my proclamation of the 22d of April, 1793, is the index to my plan. Sanctioned by your approving voice, and by that of your representatives in both houses of congress, the spirit of that measure has continually governed me, uninfluenced by any attempts to deter or divert me from it.

After deliberate examination, with the aid of the best lights I could obtain, I was well satisfied that our country, under all the circumstances of the case, had a right to take, and was bound in duty and interest to take, a neutral position. Having taken it, I determined, as far as should depend upon me, to maintain it with moderation, perseverance and firmness.

The considerations which respect the right to hold this conduct, it is not necessary on this occasion to detail. I will only observe that, according to my understanding of the matter, that right, so far from being denied by any of the belligerent powers, has been virtually admitted by all.

The duty of holding a neutral conduct may be inferred, without anything more, from the obligation which justice and humanity impose on every nation, in cases in which it is free to act, to maintain inviolate the relations of peace and amity toward other nations.

The inducements of interest for observing that conduct will best be referred to your own reflections and experience. With me, a predominant motive has been to endeavor to gain time to our country to settle and mature its yet recent institutions, and to progress without interruption to that degree of strength and constancy which is necessary to give it, humanly speaking, the mmand of its own fortunes.

Though, in reviewing the incidents of my administration, I am unconscious of intentional error, I am, nevertheless, too sensible of my defects not to think it probable that I may have committed many errors. Whatever they may be, I fervently beseech the Almighty to avert or mitigate the evils to which they may tend. I shall also carry with me the hope that my country will never cease to view them with indulgence, and that, after forty-five years of my life dedicated to its service, with an upright zeal, the faults of incompetent abilities will be consigned to oblivion, as myself must soon be to the mansions of rest.

Relying on its kindness in this, as in other things, and actuated by that fervent love toward it which is so natural to a man who views in it the native soil of himself and his progenitors for several generations, I anticipate, with pleasing expectations, that retreat in which I promise myself to realize without alloy the sweet enjoyment of partaking, in the midst of my fellow-citizens, the benign influence of good laws under a free government, the ever favorite object of my heart, and the happy reward, as I trust, of our mutual cares, labors and dangers. GEORGE WASHINGTON.

ASSASSINATION OF PRESIDENT GARFIELD.

SATURDAY, JULY 2, 1881.

At nine o'clock and thirty minutes this morning, the news flashed over the wires of an attempted assassination of the President, while waiting at the Baltimore and Potomac Depot at Washington, to take the train for New York. The party consisted of eighteen persons, principally members of the cabinet and their families. Among those who intended making the trip with the President were Mrs. Garfield, Miss May Garfield, James and Henry Garfield, Dr. Hawke, Colonel Rockwell and son, Secretary Hunt and Mrs. Hunt, Secretary Windom and Mrs. Windom, Secretary Lincoln, General Swaim and Colonel Jamison.

It was a part of the programme for General Swaim to precede the balance of the party, meet Mrs. Garfield at Long Branch and join the President and party at New York. After visiting Williams College, they were to visit the White Mountains of New England, and complete the tour about the 20th of July.

Immediately after the President entered the ladies' waiting-room with Secretary Blaine, Charles Guiteau, the assassin, within three feet of the President, fired two shots at him in quick succession and turned to flee, but was arrested by the ticket agent, Parke, and Officer Kearney, to whom he said: "I did it. I am a Stalwart. Arthur is now president."

After the President recovered from the shock and was

removed to his chamber in the White House, the following telegram was sent to his wife:

"Mrs. Garfield, Elberon, Long Branch:

"The President wishes me to say to you, for him, that he has been seriously hurt. How seriously he cannot yet say. He is himself, and hopes you will come to him soon. He sends his love to you.
"A. F. Rockwell."

The officials of the Central Railroad of New Jersey tendered Mrs. Garfield a special train over their road to Washington, which she accepted. The train consisted of a baggage and one parlor car, attached to which was one of the finest locomotives in the company's service.

Mrs. Garfield reached Washington at seven p. m. on the same day of the shooting of the President.

The sad calamity created the profoundest sensation all over the land. The wires teemed with dispatches from at home and abroad, conveying deep regrets and indignation on account of the wicked crime, and hoping for his recovery.

The following were some of the dispatches received:

ENGLAND.

"To Sir Edward Thornton, British Ambassador, Washington:

"The Queen desires that you will at once express the horror with which she has learned of the attempt upon the President's life, and her earnest hope for his recovery. Her majesty wishes for full and immediate reports as to his condition.
"Lord Granville."

SPAIN.

"Madrid, July 3, 1881.

"To the Spanish Minister, Washington:

"In the name of the King, express to the government of the United States the profound sorrow that the attempt against the President's life has caused in Spain. His majesty and his government earnestly hope for the recovery of President Garfield."

FRANCE.

"PARIS, July 3, 1881.
"M. DE GEOFROY, FRENCH MINISTER, WASHINGTON:

"Be good enough to convey to Madame Garfield the sentiment of sorrow and sympathy which the President and government feel. You will express, at the same time, to the Vice-president of the United States, the deep and profound grief which this attempt has caused throughout all France. BARTHELEMY ST. HILAIRE."

IRELAND.

"DUBLIN, July 4, 1881.
"SECRETARY BLAINE:

"In behalf of the Irish members, I beg to express our horror at the crime against the chief magistrate of the American people, and our earnest prayer that his life may be spared.
"PARNELL, House of Commons."

JAPAN.

"TOKIO, July 4, 1881.
"YOSHIDA, JAPANESE MINISTER, WASHINGTON, D. C.:

"The dispatch announcing an attempt upon the life of the President has caused here profound sorrow, and you are hereby instructed to convey, in the name of his majesty, to the government of the United States, the deepest sympathy, and hope that his recovery will be speedy. Make immediate and full report regarding the sad event.
"WOO YENO, Acting Minister of Foreign Affairs."

FROM THE KING OF ROUMANIA.

"BUCHAREST, July 4, 1881.
"To PRESIDENT GARFIELD, WASHINGTON:

"I have learned with the greatest indignation, and deplore most deeply the horrible attempt against your precious life, and beg you to accept my warmest wishes for your quick recovery.
"CHARLES."

ROME.

ROME, July 3, 1881.
Signor Mancini, Minister of Foreign Affairs, expressed condolence to Minister Marsh on behalf of the Italian government.

"GOVERNOR'S ISLAND, July 3, 1881.
"GENERAL W. T. SHERMAN, U. S. A., WASHINGTON:

"Received your last dispatch. If an opportune moment should come, please express to the President my heartfelt wishes for his complete recovery. W. S. HANCOCK."

INDIANAPOLIS, IND.

"My heartfelt sympathy is with President Garfield, and I rejoice that you think his symptoms favorable. His death would be a great public calamity. At the proper time, please communicate my kind wishes to him. WM. H. ENGLISH."

NEW YORK.

"Your 6:45 telegram is very distressing. I still hope for more favorable tidings, and ask you to keep me advised. Please do not fail to express to Mrs. Garfield my deepest sympathy.
"C. A. ARTHUR."

NEW JERSEY.

"ELBERON, N. J., July 2, 1881.
"SECRETARY LINCOLN, WASHINGTON:

"Please dispatch me the condition of the President. The news received conflicts. I hope the most favorable reports may be confirmed. Express to the President my deep sympathy and hope that he may speedily recover. U. S. GRANT."

Charles J. Guiteau, a Canadian Frenchman by birth, and the would-be assassin, deserves as little publicity as possible. For the last ten years he had been loitering about in New York, Boston and Chicago, without any apparent means or ability. Four months previous to this dastardly act he went from Chicago to Washington, and during this time made several applications to the President and Secretary Blaine, both in person and by letter, for different appointments—one to be United States Consul at Marseilles, France. Shortly after the October election, 1880, he sent the following address to the President:

" NEW YORK CITY, FIFTH AVENUE HOTEL.
"JAS. A. GARFIELD, MENTOR, O.:

"*Dear General:* I, Charles Guiteau, hereby make application for the Austrian mission. Being about to marry a wealthy and accomplished heiress of this city, we think that together we might represent this nation with dignity and grace. On the principle of first come first served, I have faith that you will give this application favorable consideration.   CHARLES GUITEAU."

Whatever induced the wretch to strike down the President, as it were, in the early morning of what promised to be a career of political usefulness, happiness and health, and a faithful devotion to the trust of a nation, remains to be seen.

The opinion of most writers is to the effect that it was the act of a crazy, maddened fanatic, while a very few consider it the result of a plot.

GUITEAU'S LETTER.

The following letter was taken from the prisoner's pocket at police headquarters:

"JULY 2, 1881.

" To THE WHITE HOUSE:

" The President's tragic death was a sad necessity, but it will unite the Republican party and save the republic. Life is a flimsy dream and it matters little when one goes. A human life is of small value. During the war thousands of brave men went down without a tear. I presume the President was a Christian, and that he will be happier in paradise than here. It will be no worse for Mrs. Garfield, dear soul, to part with her husband this way than by natural death. He is liable to go at any time, anyway. I had no ill-will toward the President. His death was a political necessity. I am a lawyer, a theologian and a politician, and I am a Stalwart of the Stalwarts. I was with General Grant and the rest of our men in New York during the canvass. I have some papers for the press, which I shall leave with Byron Andrews and his co-journalists, at 1,420 New York avenue, where all the reporters can see them. I am going to the jail.
          " CHARLES GUITEAU."

THE MODE OF SUCCESSION IN CASE OF A PRESIDENT'S DEATH.

The law, as enacted in the year 1792, remains in force, to the following effect:

§ 146. In case of removal, death, resignation or inability of both the president and vice-president of the United States, the president of the senate, or, if there is none, then the speaker of the house of representatives, for the time being, shall act as president until the disability is removed or a president elected.

§ 147. Whenever the office of president and vice-president both become vacant, the secretary of state shall forthwith cause a notification thereof to be made to the executive of every state, and shall also cause the same to be published in at least one of the newspapers printed in each state.

§ 148. The notification shall specify that electors of a president and vice-president of the United States shall be appointed or chosen in the several states, as follows: First, if there shall be the space of two months yet to ensue between the date of such notification and the first Wednesday in December then next ensuing, such notification shall specify that the electors shall be appointed or chosen within thirty-four days preceding such first Wednesday in December. Second, if there shall not be the space of two months between the date of such notification and such first Wednesday in December, and if the term for which the president and vice-president last in office were elected will not expire on the third day of March next ensuing, the notification shall specify that the electors shall be appointed or chosen within thirty-four days preceding the first Wednesday in December in the year next ensuing; but if there shall rest the space of two months between the date of such notification and the first Wednesday in December then next ensuing, and if the term for which the president and vice-president last in office were elected will expire on the third day of March next ensuing, the notification shall not specify that electors are to be appointed or chosen.

§ 149. Electors appointed or chosen upon the notification prescribed by the preceding section shall meet and give their votes upon the first Wednesday in December, specified in the notification.

IT MIGHT HAVE BEEN.

Any faction or a dissatisfied element of a political party will do more, directly or indirectly, to disorganize the party than its greatest common foe can do.

When two or more great parties meet in a political canvass they measure their strength at the ballot-box, and no matter how bitter the strife, a redeeming quality is found in the fact that every American citizen acquiesces peacefully in the result. The minority yields to the majority and our chief magistrate is no longer an exponent of some political organization, but the acknowledged ruler of all. This fact has been fully demonstrated since the attempt upon the life of President Garfield. Telegrams of condolence poured in from all the people of our country, regardless of all party affiliations. And the nations of Europe manifested a deep sympathy in our national sorrow.

The bitter animosity engendered at a political warfare subsides at the close of an election, and quiet prevails until occasion calls forth another strife for political supremacy.

Not so with a *faction* of a party. It does not always end at an election, and if not thwarted in time, will not only defeat itself, but utterly demoralize and ruin the very party of which it is a factor.

Washington says, in his " Farewell Address," that the alternate domination of one faction over another, sharpened by the spirit of revenge natural to party dissension, which, in different ages and countries, has perpetrated the most horrid enormities, is itself a frightful despotism.

The disorders and miseries which result gradually

incline the minds of men to seek security and repose in the absolute power of an individual, and, sooner or later, the chief of some prevailing faction, more able or more fortunate than his competitors, turns this disposition to the purposes of his *own* elevation on the ruins of the public liberty. Toward the preservation of your government and the permanency of your present happy state, it is requisite, not only that you steadily discountenance irregular opposition to its *acknowledged authority*, but also that you resist with care the spirit of innovation upon its principles, however specious the pretexts.

In the present crisis, with our President badly wounded, we have the outgrowth of political discord.

While we do not entertain the idea that the attempted assassination was a conspiracy, we believe that the wicked man's mind, if deranged at all, drifted into the current of public agitation and imagined that the welfare of the country demanded the life of the President at his hands, whereby had there been no discontent in the party—had the administration been fully indorsed and assisted by the united effort of all its members, the assassin's mind might have taken some other course, and, if ending in murder at all, it could not have been more disastrous, while it might have been of much less consequence.

Suppose the attack had resulted in instant death, as was the intention, and the death of Arthur had followed immediately, from whatever cause, then we should have been without a president, and the field open for an election.

Each congress elects its own speaker. The forty-sixth congress went out of power March 4, 1881, and the next, or forty-seventh congress, would not convene in regular session until December, 1881. The United States senate adjourned in May, 1881, without electing a president of the senate pro tem. Now, in the absence of the president and vice-president, the president of the senate pro tem, or, if there be none, then the speaker of the house, shall act until a president is elected.

In this case and at this time, we have neither a president pro tem of the United States senate nor a speaker of the house of representatives. Consequently we would be without a president until one could be elected by the people. And the length of this interregnum would depend upon circumstances.

By section 147, the secretary of state shall prepare for an election.

By section 148, if there should be a space of two months between the secretary's notification of an election and the first Wednesday in December following, then we would be without a ruler for two months; but if there should not be a space of two months, etc., then we would be without a president until the second December following said notification; provided the term of the president and vice-president would not expire at the March ensuing the first December, following the secretary's notification of an election. And in the present case it would not.

But this is merely a speculation on the life and death of two men under the present circumstances. While an interregnum is possible at any time, the contingency in the present case was almost tangible, and it is to be hoped will never occur again in the history of our nation.

$65. Beatty's Organs.

Warranted Six Years.

18 BEAUTIFUL STOPS

Including the Famous **Vox Celeste, French Horn Solo Combination, Vox Humana, Sub Bass, and Oct. Coupler.**

Shipped on test trial to all parts of the world.

ORDER ONE.

The "London,"

New Style No. 5,000.

(1) Diapason Forte. (2) **SUB-BASS.** (3) Principal Forte. (4) Dulcet. (5) Diapason. (6) **Grand Organ.** (7) VOX HUMANA. (8) Æolian. (9) Echo. (10) Dulciana. (11) Clarionet. (12) **Vox Celeste.** (13) **Octave Coupler.** (14) Flute Forte. (15) **French Horn Solo.** (16) Bourdon. (17) **Grand Knee Stop.** (18) Right Knee Stop.

(19) *BEATTY'S PATENT STOP ACTION.*

SPECIAL NOTICE.—On and after April 1st, till Sept. 30, 1881, "London" New Style No. 5,000, will contain 5 full sets reeds, 18 stops (as described herewith) instead of 4 sets, 14 stops as formerly. This is a special offer and made solely as a special Mid-Summer Holiday offer. **I DEFY Competition.**

It contains five octaves, five full sets of the **Celebrated Golden Tongue Reeds,** as follows: 2 sets of 2½ Oct. each (regulars), 1 set powerful SUB-BASS, 1 set FRENCH HORN, also 1 set Vox Celeste.

☞ **5 FULL SETS IN ALL,** two knee stops, handles, lamp stands, pocket for music, **Solid Black Walnut Case,** carved, veneered, extra **Large Fancy Top,** as shown in the accompanying picture. It is a very **STYLISH CASE.** Upright bellows, steel springs, metal foot plates, rollers for moving, etc.

Height, 72 in.; Depth, 24 in.; Length, 46 in.; Weight, 400 lbs.

☞ Price, boxed, delivered on cars here, with stool, book, music, only **$65.** It is **UNEQUALED.**

☞ Organs, Church, Chapel and Parlor, $30 to $1000, 2 to 32 Stops. Baby Organs, only $15. Splendid New Organs, 4 full set reeds, for $48. The "Paris" offered for $85, a magnificent instrument. Other desirable New Styles now ready. **PIANOS,** Grand Square and Upright, $125 to $1,600.

☞ **REMIT** by Post-Office Money Orders, Express prepaid, Bank Draft, or by a Registered Letter. **Money Refunded** and freight charges will be paid by me both ways, if the instrument is not *just as represented.* ☞ Please send reference if you do not remit with order. *Be sure* to send for *Latest Illustrated Catalogue,* "*Beatty's Quarterly,*" *Midsummer Holiday Circulars,* etc., sent free.

Address or call upon **DANIEL F. BEATTY,** Washington, New Jersey.

C. W. BROWN,

PLAIN AND FANCY

Job Printer,

Binder, Blank Book Manufacturer,

AND PAPER BOX MAKER.

Cor. 5th and Ohio Streets,

TERRE HAUTE, IND.

Philip Schloss,

THE POPULAR

Merchant Tailor

AND CLOTHIER,

420 Main Street,

TERRE HAUTE, INDIANA.

(North Side, bet. Fourth and Fifth Sts.)

MOORE & LANGEN,

ENGLISH AND GERMAN

Book and Job Printers

16 South Fifth Street,

TERRE HAUTE, INDIANA.

Daily Express Building.

H. F. SCHMIDT.   J. BERNHARDT.

H. F. SCHMIDT & CO.,

DEALERS IN

Watches, Clocks, Jewelry,

Silverware and Spectacles,

403 Main St., opp. Opera House,

TERRE HAUTE, IND.

Repairing Promptly Attended to. All Work Warranted.

C. EPPERT,

Photographer

No. 323½ Main St.,

Between 3d and 4th,

TERRE HAUTE, IND.

J. H. SYKES,

Wholesale and Retail Dealer in

Hats, Caps, Furs,

AND STRAW GOODS.

419 Main Street,

TERRE HAUTE, IND.

Sheldon Swope,

DEALER IN

Watches, Clocks,

JEWELRY AND SILVERWARE,

Opera House Block,

TERRE HAUTE, IND.

Repairing Promptly Done. All Work Warranted.

Townley Bros.,

Wholesale and Retail

STOVES, TINWARE

And House-Furnishing Goods,

514 Main Street,

TERRE HAUTE, IND.

Central Book Store	NOYES ANDREWS & SON,
J. Q. BUTTON & CO.,	DEALERS IN
Booksellers and Stationers,	Men's, Ladies', Misses' and Children's
524 Main Street,	FINE SHOES,
TERRE HAUTE, IND.	505 Main Street,
	TERRE HAUTE, IND.

IF YOU WANT
China, Glass, Queensware
LAMPS, CHANDELIERS, CUTLERY, FRUIT JARS, ETC.,
AT WHOLESALE OR RETAIL,
Call on **H. S. RICHARDSON & CO.**
307 Main Street, TERRE HAUTE, IND.

THIS BOOK

Will be sent POST-PAID to any part of the United States, upon receipt of

One Dollar and Fifty Cents.

Send Money by Registered Letter, at our risk.

Call on or Address,

T. W. FUQUA,
TERRE HAUTE,
Cor. 6th and Cherry Sts.   IND.

www.ingramcontent.com/pod-product-compliance
Lightning Source LLC
Chambersburg PA
CBHW031817220426
43662CB00007B/691